The Palisades Sill & Watchung Mountains:
A Tale of Pangea Rifting in the Newark Basin

By William A. Szary

Copyright 2019. Earth2Energy. All Rights Reserved.

Book Cover: Front- Great Falls over the Watchung Basalt in Paterson New Jersey by Fine Art America. Back – Palisades Sill along the Hudson River, west side in Palisades Interstate Park by Palisades Interstate Park.

Library of Congress Catalog in Publications Data:

Szary, William A. The Palisades Sill & Watchung Mountains:
A Tale of Pangea Rifting in the Newark Basin

Includes references

ISBN 13: 9781711758091

Earth2Energy Educational Publishing
Port Richey FL 34668

Earth2Energy is a Registered Trademark

Table of Contents

Part I. The Palisades Sill 5

Geography and Geology Summary
Introduction
Geologic History of the Palisades Sill
Trace Element Geochemistry
Palisade Sill Discussion of Geochemistry
Discussion of Trace Element Geochemistry
Palisades Sill Plume Source
Conclusions

Part II. The Watchung Mountains 17

Geography and Geology Summary
Mineralogy
Geologic Setting of the Watchung Basalt Flows
Location and Physiography of the Watchungs
Geologic Setting
Initial Graben Subsidence
Longitudinal Crustal Warping
Second Generation Graben Subsidence
Fragmentation of Transverse Warped Structures
Final Episode of Tectonic Activity
Structural Characteristics
Watchung Glaciation
Stratigraphy and Geologic Age
Local Stratigraphic Relations
Upper Triassic Fissure Eruptions in New Jersey
Igneous Activity Geologic Age
Regional Stratigraphic Relations
The Nature and Products of Basalt Eruptions
Geologic Processes Responsible for Jointing Systems
Petrochemistry
Megascopic Characteristics of the Basalt
Probable Character of Pristine Gases
History of Tectonic Subsidence and Volcanism in the Newark Basin
Structure of the Newark Basin at the close of the Triassic Period
Structure of the Newark Basin after post-Triassic Deformation
Regional Arch Formation
Triassic Tensional History
Development Effects of the Newark Basin

Supplementary Topics
Fissure Eruption Characteristics
Bubble Formation Processes and Movement
Natural lava Flow Bubble Formation

References 68

Part I. The Palisades Sill

Geography and Geology Summary

The following extract was obtained from Wikipedia (2019) as an introduction to the Palisades Sill.

The Palisades Sill is a Triassic 200 million year old diabase intrusion. It extends through portions of New York and New Jersey. It is most noteworthy for the Palisades, the cliffs that rise steeply above the western banks of the Hudson River.

The outcrop of the Palisades Sill is quite recognizable for its prominent cliffs above the Hudson River. It is easily seen from the western portions of Manhattan. The exposure is approximately 80 km (50 miles) long, most of it following the Hudson River. It first emerges in Staten Island, New York City. The sill then crosses the state line into New Jersey where Jersey City, Union City, Fort Lee, and Englewood Cliffs all lie on it.

The sill eventually crosses back into New York, following the Hudson River north until reaching Haverstraw. It is at this point that the sill makes a turn to the west, where it disappears near Pomona. At this turn, the sill cuts across local strata, making it a dike in that area, not a sill.

It has been proposed that the sill re-emerges in two locations in Pennsylvania where the outcrops are also discordant with local strata but this idea is not generally agreed upon, and discussions of the Palisades Sill are usually limited to the exposure in New Jersey and New York.

A portion of the sill is also home to the Palisades Interstate Parkway, a stretch of road that passes through the park area preserved to protect its natural history (**Figure 1**).

Figure 1. Upper left- extent of the Palisades Sill beginning in Staten Island, extending through northeastern New Jersey into New York State (red line). Upper right- the Palisades Sill overlooking the Hudson River facing south. Lower Left- USGS diagrammed representation of the Newark Basin sills along Interstate 80 in New Jersey. Lower right – columnar jointing example in the Palisades Sill as described in the text below. Source: Wikipedia, no published date, posted on the internet.

The end of the Triassic Period saw large scale rifting during the break up of Pangea. What is now eastern North America began to separate from what is now northwestern Africa, creating the young Atlantic Ocean. Magma was generated through decompression melting, and a portion of it was intruded into the sandstones and arkoses of the Stockton Formation within the Newark Basin, of the Eastern North America Rift Basins. The magma would eventually solidify and, after millions of years, the overlying rocks would be uplifted and eroded exposing the Palisades Sill as we know it today.

The composition of the sill is that of diabase, although its mineral assemblage is not uniform throughout the thickness of the body. The mineralogy of the sill consists principally of plagioclase feldspar, several varieties of pyroxenes, and olivine with minor biotite, titanite, zircon, and oxides.

Most researchers report that the sill becomes progressively differentiated as one moves away from either the upper or lower contact. The sandwiched horizon is the term given to the central region where both cooling fronts met. It is here where the diabase is the most differentiated.

The most intriguing geological feature of the sill is 10 meter thick olivine rich zone roughly 10 meters (30 feet) from the lower contact. The modal percent of olivine goes from 0 to 2% within the main body of the sill to up to 28% within this layer. It is the origin of this layer, and subsequently the sill as a whole, that generated much of the attention, as well as varying origin theories proposed for the intrusion.

There is an average stratigraphic thickness of 300 meters (~1000 feet) with the famous cliffs rising 100 meters (300 feet) on average above sea level. The intrusion dips between 10 and 15 degrees westward for most of its length. It has been determined through stratigraphic studies that the sill was intruded at a depth of approximately 3 to 5 km. These studies also concluded that the sill was emplaced in a position nearly identical to its current one (10 to 15 degree dip) that is further confirmed by the sill vertical orientation of the columnar jointing.

It has been proposed that the Watchung Basalt flows of the Watchung Mountains are extrusive eruptions of the same magmas that created the Palisades Sill. Magmatic and gravity measurements have indicated the presence of a large subsurface dike between the Palisades intrusion and the Ladentown Basalt, an extrusive body of Watchung basalt north of Suffern, New York. More recently, the various Watchung flows have been correlated to geochemically distinct layers within the Palisades Sill, bolstering the theory that eruptions of the Palisades margins were responsible for the episodic flood basalts of the Newark Basin.

Due to the presence of the olivine rich zone, the normally difficult task of determining the history of an igneous body becomes event tougher.

The original studies concluded that the sill was the result of a single injection of magma. The variety in mineralogy was credited to simple crystal fractionation.

The next model introduced whole rock geochemistry data and determined that there were at least two separate injections; an olivine rich magma was followed by normal tholeiitic basalt.

The third hypothesis proposed at least three but probably four separate pulses with the olivine rich magma being the final one. This was also conjectured through the use of whole rock chemistry.

Introduction

A Chapter 5 of an unknown publication was posted on the internet without reference information. The following extraction is from that undocumented publication.

The extensively studied 200 mya Central Atlantic Magmatic Province (CAMP) is considered to be the biggest Large Igneous Province (LIP) on the planet covering up to 7×10^4 km^2. This igneous province has been linked to the early Mesozoic initial opening of the Central Atlantic Ocean. The opening of the Central Atlantic Ocean fragmented the CAMP into several segments that occur on four different tectonic plates today. This magmatic event has been compared to formation of flood basalt provinces such as the Siberian and Deccan Traps, in that each may be genetically linked to a global faunal extinction. For CAMP this is the event recorded at the Triassic-Jurassic boundary. Although many authors have done extensive work on CAMP related tectonics and magmatic provinces, there is still no consensus on its origin, and many aspects of the CAMP remain controversial.

The CAMP related LIP is different from others in that it constitutes almost entirely of dikes and sills with scarce volcanic outflows. In the pre-Atlantic Ocean reconstruction, this dike swarm defines an overall radiating pattern extending nearly 300 km from its focal point. Although the distribution of dikes on a local scale shows more complex patterns, on a large scale it represents the best example of a complete radiating dike swarm system comparable only with some of the radial dike swarms of Venus.

Extensive isotopic dating has shown this even to be of short duration with magmatism in all regions occurring within a few million years at 200 mya. Although non-plume models have been considered for the CAMP event, a mantle plume appears to be necessary to explain the radiating dike pattern and the generation of such a huge area of basaltic magmatism within only a few million years (**Figure 2**).

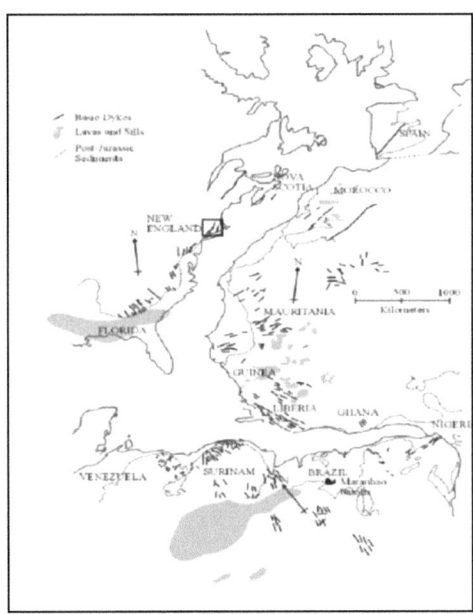

Figure 2. General distribution of Early Jurassic tholeiitic dikes, lavas, and sills in a pre-drift reconstruction at 200 mya. The square represents the Palisades Sill region.

The study of regionally extensive dike and sill systems represents one of the fundamental tools in the analysis of LIPs to classify their origin and geodynamic setting. In this context, the detailed geochemical study of the Palisade Sill reported here may shed light on some of the unsolved questions linking LIPs and regional dike swarms such as 1) the association of radiating dikes with either sub-lithospheric plume impingement or mantle insulation beneath highly refractory cratons and 2) vertical emplacement versus far reaching (>3000 km) and nearly instantaneous (in less than a few million years) lateral migration of magma from its source.

It is generally agreed that the Palisades Sill basalts are an expression of the CAMP Magmatism that related to the earliest stages of the opening of the Central Atlantic. In terms of trace elements and multiple element isotopic systematics, this poorly studied area of CAMP, presently located in northeastern America is similar to the voluminous low titanium tholeiites associated with CAMP and distinctly different from the smaller volumes of high titanium Magmatism related to the CAMP.

Geologic History of the Palisades Sill

The initial fragmentation of the Pangea supercontinent was accompanied by extensive tholeiitic Magmatism now represented by sills, dikes, and minor lava flows in four continents along both sides of the Central Atlantic Ocean, on the eastern margin of North America between Nova Scotia and Florida, Western Europe (Spain and France), West Africa (Morocco to the Ivory Coast), and northern South America (French Guyana, Surinam, and Brazil).

The Palisades Sill is one of several diabase intrusions located in the system of Eastern North America rift basins formed by significant crustal extension during the break up of Pangea during the Late Triassic. This early Jurassic Palisade Sill that intruded into the Triassic continental sedimentary rocks of the Newark Basin has been cited as the classic example of a vertically differentiated sill. The Palisades Sill and other sills in the Newark Basin are used as markers to define the Triassic-Jurassic boundary. The intrusion is an approximately 200 mya, 300 meter thick diabase sill intruded into sandstones and arkoses of the Newark Basin. The outcrop extends for 80 km in a north-south direction from central Staten Island in New York through a significant portion of New Jersey, along the western bank of the Hudson River. Here, it turns westward and is discordant with the local strata and hence referred to as a dike locally. It may connect under cover with the Rocky Hill and Lambertville Sills to the south for a total strike length of 150 kms. Field relations and petrography have been described by a number of authors (**Figure 3**).

Several authors have done major element analyses of the various layers of the Palisade Sill to understand compaction and differentiation processes experienced by this sill after emplacement. However, very limited trace element and isotopic geochemistry related work has been done in this region.

Trace Element Geochemistry

The basalts and gabbros show uniform patterns with slight light rare earth element (LREE) enrichment and a relatively gentler slope for the heavy rare earth elements (HREE). In contrast to the basalts and gabbros, sandstones and chilled margins show a much wider range of chondrite normalized REE patterns, especially LREEs. These sandstones and chilled margin basalts show LREE enrichment, 20 to 100 times that of chrondrite; HREE 10 to 40 times that of chrondrite, and strong negative europium anomalies.

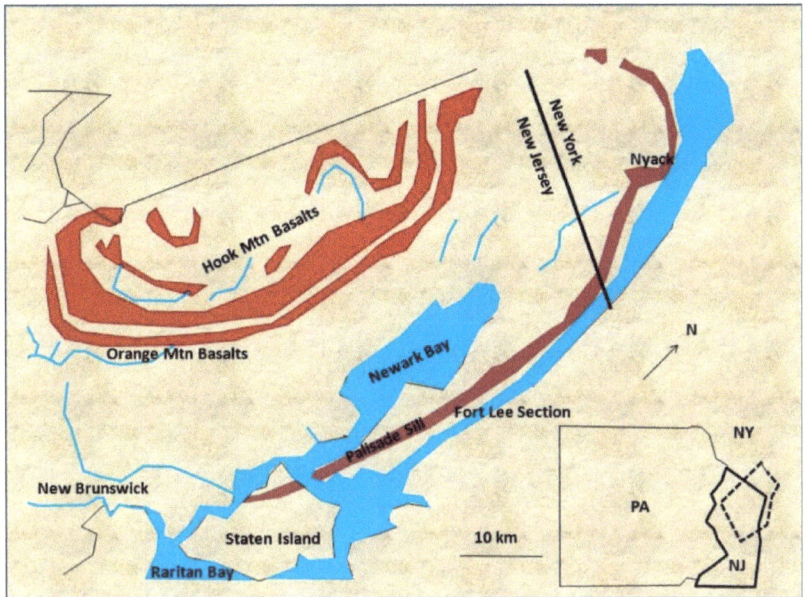

Figure 3. Distribution of early Jurassic igneous rocks throughout the northern Newark Basin. The location of the Palisades sill basalts is shown in the side map on the bottom right.

Twenty two compatible and incompatible trace element concentrations patterns for the Palisades Sill lavas and associated sandstones and chilled margin basalts are shown normalized to primitive mantle. Primitive mantle normalized basalts and gabbros show low barium, high uranium, lead, zinc, and hafnium concentrations, mildly negative niobium-tantalum anomalies, and gently sloping HREEs. In general, basalts and gabbros of the Palisades Sill are similar to and 30 times more enriched than primitive mantle.

Sandstones and chilled margin basalts have enrichments and depletions in the same elements as the Palisade Sill basalts and gabbros, but the magnitude of these enrichments and depletions are much larger for the sandstones and chilled margin basalts. In addition, they also have negative strontium and positive thorium peaks. The niobium-tantalum anomaly is much stronger in the sandstones than in the chilled margin basalts.

The lanthanum-barium and lanthanum-niobium ratios of the Palisades Sill data as well as associated sandstones and chilled margins were compared to various mantle reservoirs along with CAMP data from France, Guinea, and Guyana.

There is a close correspondence of the Palisades basalts and gabbros with oceanic island basalts (OIBs), and low titanium CAMP related tholeiites from Guinea and France. Peridotite melting models are based on composition of olivine (60%), orthopyroxene (25%), clinopyroxene (10%), and garnet (5%).

Palisade Sill Discussion of Geochemistry

The participation of a mantle plume as the main cause of the CAMP is a matter of intense debate and there are many arguments that can be put forward to either support or reject the participation of a Central Atlantic Plume (CAP) in the petro-genesis of the primary magmas. In this section, an attempt is made to identify either a plume or a sub-continental mantle lithosphere source for the Palisade Sill basalts as well as other low titanium CAMP tholeiites based on geochemistry.

Discussion of Trace Element Geochemistry

The Chondrite normalized REE patterns of the basalts and gabbros from the Palisades Sill display only minor variability along the entire sill. There is slight LREE enrichment in these rocks which is typical of both high and low titanium CAMP tholeiites. In particular, the REE patterns of the Palisade Sill basalts strongly resemble the low titanium tholeiites from Europe and Brazil. The Palisades Sill basalts as well as other high and low titanium CAMP tholeiites are distinctly different from Central Atlantic MORB in their REE patterns and are likely derived from an enriched mantle. In contrast to the restricted REE patterns of the Palisade basalts, the sandstones and chilled margins have a much wider range of REEs with a strong negative europium anomaly that is characteristic of continental crust. The sandstones are the likely contaminants of the Palisade Sill basalts. The chilled margins that formed by the sudden cooling of the intrusive lava with the surrounding rock have REE patterns similar to the sandstones, and likely experience nearly 100% contamination by the sandstones.

On the multi-element primitive mantle normalized diagram, the Palisade basalts are flat with the exception of low barium and very high lead. There are small negative niobium-tantalum and positive zirconium-hafnium anomalies in a few samples. Tholeiites from French Guyana and Spain also show similar patterns.

The nearly flat HREE pattern implies an absence of garnet in the source, the depth of melting for these rocks may have been in the field of stability of spinel peridotite. These rocks were possibly derived from a mantle 20 to 50 times more enriched than the primitive mantle. The sandstones and chilled margins have low barium, strontium, and europium, negative niobium-tantalium anomalies, and high lead when normalized to the primitive mantle. Notice that both Palisade Sill as well as sandstones show the enrichments and depletions in the same element, but the magnitude of enrichment and depletion is higher in the sandstones compared to the Palisade basalts. This indicates that the Palisade Sill lavas to be contaminated by these sandstones prior to or during emplacement. The chilled margins experienced very large degrees of contamination by the sandstones and hence mimic their trace element patterns.

Most continental flood basalts (CFBs) are characterized by both high and low titanium basalts. The Palisades Sill basalts appear to be derived from a source region similar to other low titanium CAMP as well as other continental flood basalts. These basalts have similar lanthanum to barium ratios and slightly higher lanthanum to niobium ratios compared to ocean island basalts. They are distinctly different from the high titanium CAMP tholeiites from Guyana which lie entirely within the ocean island basalt field and have relatively higher lanthanum to barium ratios.

The systematic differences between the high and low titanium CAMP tholeiites may indicate that these two types of tholeiites have different mantle sources. However, it is also possible that both high and low titanium tholeiites are derived from an enriched mantle and the low titanium tholeiites have been contaminated by the continental lithosphere whereas the high titanium tholeiites lie close to the primitive mantle and may or may not have asthenospheric contamination. Another possibility is that the high and low titanium lavas of the CAMP provinces may have been derived by different depths and degrees of melting of the same mantle.

Studies of niobium-Tantalium variation in MORBs, komatiites, depleted mantle xenoliths, and chondritic meteorites have suggested that the niobium to tantalium ratio of the Earth's mantle has a chrondritic value of 17.5 and that there is no significant fractionation of these elements, at least to a large degree of partial melting. However, experimental studies of the partitioning of these elements between various melt compositions and potential residual phases in the mantle suggest that in certain circumstances this pair of geochemically similar elements may be fractionated. These effects have been modeled assuming primitive niobium and tantalium concentrations, and the calculated likely maximum variation of niobium concentration with the niobium to tantalium ratio in the melt and the residue at various degrees of batch melting.

At very low degrees of melting, niobium to tantalium ratio values may be similar to leucites, but calculated niobium concentrations are much higher, and zircon to niobium ratio values for the melt are much lower due to significant reduction of niobium to tantalium ratio in the residue after removal of a very small melt fraction. It can be inferred that the low titanium Palisade Sill basalts were derived from 15% melting of a slightly depleted peridotite. The high titanium tholeiites may have been products of the initial small melt fractions that left behind a slightly depleted peridotite source for the low titanium CAMP tholeiites.

The Palisade Sill basalts and most of the CAMP investigated throughout the four circum-Atlantic continents are typically tholeiitic low titanium continental flood basalts which differ fundamentally from MORB by higher concentrations of LREE. On the basis of trace element data discussed above, the low titanium Palisade Sill basalts appear to have been derived by melting of a spinel lherzolite and are likely contaminated by the continental crust represented by the sandstones and chilled margin basalts.

Palisade Sill Plume Source

The relative enrichment of these elements in CAMP has long been debated and explained by different mechanisms. 1) Derivation from an enriched sub-continental lithosphere mantle source with no or limited crustal contamination during magma ascent; 2) derivation from an asthenospheric MORB like source with a more significant crustal contamination, and 3) derivation from an incipient plume head.

Geochemical data shows no evidence of a depleted asthenospheric MORB source for the CAMP derived dikes and sills. Derivation from an enriched sub-continental lithosphere requires lithospheric thinning to drive decompressional melting which is unlikely to produce effusive, rapid volcanism. In plume models, mantle material originates at or near the core mantle boundary and rises buoyantly through the mantle, spreading laterally as the plume head encounters the thin eroded lithosphere and produces voluminous melts. Is it possible that all the CAMP (6000 km) is a result of a single process melting of a heterogeneous lithospheric mantle initiated by plume impingement? A more simple answer is to attribute the CAMP to a hotspot system. Geochronological data show that most parts of the investigation of CAMP were active in the early Jurassic (200 mya). This time coincidence of the CAMP age argues for one plume for the CAMP magmatism. The brief and extremely widespread tholeiitic magmatism associated with the CAMP implies that an anomalously hot mantle extended over a very wide area and melted extensively. In general, geochemical and geochronological data on CAMP are consistent with models that suggest that an upwelling plume head separated from the plume tail and that the plume material spread over a very large area by ambient mantle flux.

The CAMP can be compared to the East African Rift System (EARS) which is a classic example of ongoing continental rifting and provides an excellent framework to investigate magmatism in an extensional setting. The EARS rift system extends over 4000 km from the Red Sea in the north to Mozambique in the south representing a 150 km wide zone of NW-SE trending extension. The seismically and volcanically active EARS is the youngest mantle plume province worldwide with one or more upwellings impinging on thick cratonic lithosphere since 45 mya caused by an anomalously hot asthenosphere. Discrete rifting episodes have recently been observed in the Afar triple junction in Ethiopia. In this region, batches of molten mantle rocks have risen into cracks and fractures to form long thin vertical sheets of new crust in the form of dikes, often feeding surface eruptions of basalts. These dikes serve to transport melt percolating upward from mantle source zones and also had accumulated in magma chambers or thin horizontal magma sheets within the crust. The dikes, along with faults constitute plate boundary separation within the crust in this region. Tomography and seismic images, numerical models, and geochemical data collectively support a single plume with multiple stems originating in the African super plume.

The CAMP was derived from a compositionally heterogeneous super plume similar to the present day EARS with multiple stems acting as feeders over a large aerial extent.

Conclusions

In this study, the Palisade Sill basalts and gabbros were correlated to other low titanium CAMP related magmatism from eastern North America, South America, Europe, West Africa, and Canada by their geochemical and niobium, strontium, and lead isotopic signatures. When compared to the small volume high titanium CAMP related magmatism, the low titanium CAMP lavas have a more radiogenic strontium and less radiogenic niobium component as well as higher values of lead. Based on the geochemical data presented, the Palisades Sill basalts were derived from slightly enriched oceanic island basalt like mantle source. Further, these rocks were derived from 15% melting of a slightly depleted spinel peridotite. Since other low titanium CAMP lavas have similar geochemistry as well as eruption ages as the Palisades Sill basalts, it is safe to assume the same source for these tholeiites across the four continents where they were emplaced.

Although the collective trace element and niobium strontium and lead isotopic signatures of low titanium CAMP magmatism cannot unequivocally discriminate between a heterogeneous sub-continental lithosphere and a plume source contaminated by the continental lithosphere, the argument for the latter is based on the short time of eruption of all these lavas as well as the absence of an electromagnetic like component in the low titanium lavas.

It is suggested that the CAMP was derived from a compositionally heterogeneous super plume similar to the present day East African Rift System with multiple stems acting as feeders over a large areal extent. Compositional heterogeneity may have been caused due to variability in lithosphere=asthenosphere boundary, or due to different degrees and depths of melting as has been suggested for the heterogeneous basalts derived from the EARS.

Part II. The Watchung Mountains

Geography and Geology Summary

The following introduction was posted on the internet by Wikipedia (2019).

The Watchung Mountains, once called the Blue Hills are a group of three long low ridges of volcanic origin between 400 and 500 feet high lying parallel to each other in northern New Jersey in the United States. The two most prominent ridges are known as the First Watchung Mountain (the southeastern ridge), the Second Watchung Mountain (the northwestern ridge) stretch for over forty miles from Somerville (Somerset County) in the southwest to Morris County, Union County, Essex County, and Passaic County to Mahwah in Bergen County in the northeast. The less prominent and discontinuous ridge formed by Long Hill, Riker Hill, Hook Mountain, and Packanack Mountains sometimes referred to as the Third Watchung Mountain located on the northwest side of the Second Watchung Mountain.

The First Watchung Mountain and Second Watchung Mountain are often erroneously referred to as Orange Mountain and Preakness Mountain. Historically, the names Orange and Preakness were applied to sections of these ridges. The confusion arose from the fact that the First Watchung Mountain is said to be composed of Orange Mountain Basalt, while the Second Watchung Mountain is composed of Preakness Mountain basalt. Names applied to the basalts are actually type locations, that is that the rock found at Orange Mountain is exclusive to all of First Watchung Mountain, while the type of rock found at Preakness Mountain is exclusive to all of the Second Watchung Mountain. Like First and Second Watchung Mountain, Third Watchung Mountain is sometimes referred to its type locality. The entire length is referred to as Hook Mountain on some occasions.

In addition to the three main ridges of the Watchungs, a smaller fourth ridge exists south of Morristown and west of the Third Watchung Mountain. While attaining elevations of over 400 feet above sea level, the ridge lacks topographic prominence, only rising to about 100 feet above the surrounding terrain. Only one portion of the ridge is named, a southern section underlying Harding Township known as Lees Hill.

All of the ridges lie to the east of the higher Appalachian Mountains which in northern New Jersey are often referred to as the New York-New Jersey Highlands. Together with the Appalachian Mountains to the west, the Watchungs pen in an area formerly occupied by the prehistoric Glacial Lake Passaic. The Great Swamp, a large portion of which is designated as the Great Swamp National Wildlife Refuge, is a remnant of the lake, presently retained by the Third Watchung Mountain.

About 200 million years ago, magma intruded into the Newark Basin, then an active rift basin associated with the breakup of the supercontinent called Pangea. The magma was initially contained within the sedimentary strata of the basin, forming large intrusions like the Palisades Sill, but it ultimately broke out at the surface through large episodic eruptions. The Watchung Mountains were originally formed from these eruptions, consisting of these separate flood basalts that may have filled nearly the entire Newark Basin. Each time the basin filled with basalt, which cooled into blocky trap rock, a period of limited volcanic activity followed, allowing sediment to be deposited on top of the previously erupted layer of basalt. In this way, the Newark Basin became layered with alternating strata of Watchung basalt and Jurassic sedimentary rock (**Figure 4**).

Throughout the Early Jurassic, the Newark Basin underwent extensive dipping and folding. The western side of the basin plunged deeper into the crust, tilting the basins' strata to an angle between 5 and 25 degrees. Localized deformation of the western edge of the basin along the Ramapo Fault System formed alternating synclines and anticlines that warped the layers of basalt and sedimentary rock (**Figure 5**).

Erosion began to attack the basin as rifting failed and deposition of new sediments ceased. Over millions of years, erosion occurred downward through tilted rock of the basin, eventually encountering the basalt layers, significantly more resistant to erosion that the surrounding sedimentary rock. The result of this erosion exposed the edges of the eroding basalt layers which persisted much longer than the exposed sedimentary rock layers resulting in a projection above the surrounding surface terrain as high ridges.

Figure 4. The breakup of Pangea created a rift zone along the North American margin when Africa began to pull away during the initial stages of Pangea's destruction. New Jersey was flooded by the opening of the beginning stages of the Atlantic Ocean.

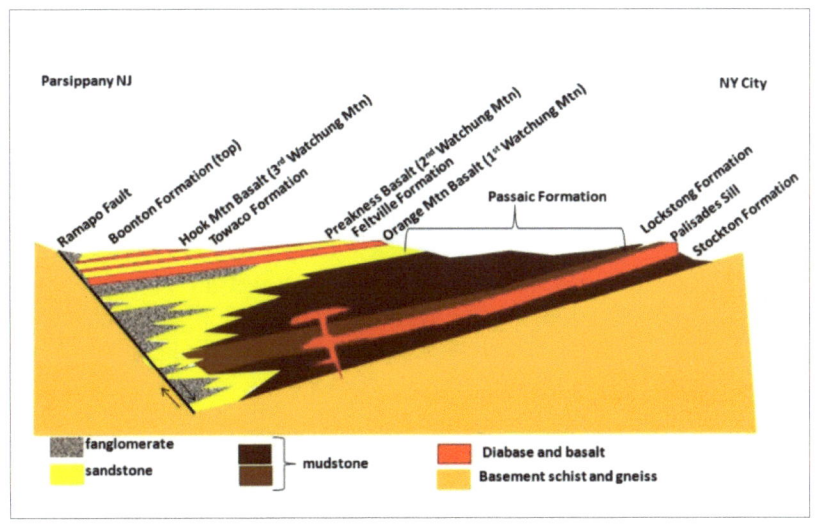

Figure 5. Cross section showing the Newark Basin and the Watchung Mountain intrusions into Jurassic sedimentary rock. Source: USGS.

Today's landscape exposes the flood basalts which are preserved in the synclines adjacent to the Ramapo Fault System. In these synclines, the basalt layers are thick and warped into down ward dipping trap rock sheets, descending below the current erosional surface of the basin. Notably, the syncline preserved not only the basalt layers, but also some overlying Jurassic sedimentary rock. The largest syncline in the basin, the Watchung Syncline, contains the greater portion of the Watchung flood basalts as they appear today.

The protecting, eroding edges of the flood basalts preserved in the syncline form the three ridges of the Watchung Mountains. Jurassic sedimentary rock layers between and above the ridges form the Feltville, Towaco, and Boonton Formations. Elsewhere in the Newark Basin, smaller synclines preserve the Watchung Outliers, additional fragments of the flood basalts and associated overlying sediments that have survived into the modern era.

Because the majority of the Watchung Mountains are composed of extrusive igneous trap rock, they display characteristic columnar jointing and stacked lava flows. These features are readily noted along the eastern facies of the ridges, which often present several principles or vertical escarpments. Similar features can be seen in the Palisades Sill, although these were formed within the Earth's crust. Additionally, the Watchung's feature not only blocky aa lava, but also ropey and billowing pahoehoe flows.

The magma which generated the Watchungs and the Palisades also formed the intrusive igneous Sourland Mountain in Central New Jersey, as well as a series of smaller outlying volcanic ridges in the region. Cushetunk Mountain, a ring shaped volcanic mountain between Sourland Mountain and the Watchungs is of the same geologic lineage.

The Metacomet Mountains in the Connecticut River Basin, another aborted rift valley, came into existence around the same time as the Watchungs, also through extrusive eruptions. While noncontiguous, the two ranges may be considered geologic cousins, forming under similar circumstances during rifting of Pangea. The same erosive and tectonic forces which elevated the Watchungs also served to raise the Metacomets.

Mineralogy

The Watchungs are composed principally of volcanic basalt which historically has been used in railroad beds and road construction. In addition, in many places the mountains are underlain by red and white sandstone which is also used in building construction. Mica and calcareous spar often accompany these sandstone beds.

Due to the volcanic nature of the Watchungs, zeolites including prehnite, analcine, and stilbite were formed from reaction of mafic rocks in alkaline environments, are found along ridge lines. Agate, primarily in the form of chalcedony, and crystalline quartz (sometime in the form of amethyst) are prominent in the ancient lava flows of the Watchungs and are typically seen as embedded nodules along exposed fronts. Datolite, another nodular mineral, has been found embedded in the volcanic rock around the Great Falls. Additionally, jasper, and satin spar are known to exist with the northwestern Preakness Range.

Copper also can be found in the Watchungs. Near Belleville, ore containing 8% copper was discovered, and a copper mine once operated in this area. Other copper bearing ores have been noted near Paterson. These ores typically contain cuprite (red copper ore) and/or copper carbonate in a matrix of red or gray sandstone. Pyritous copper, also known as chalcopyrite is not known to exist in ores found in the Watchungs.

Geologic Setting of the Watchung Basalt Flows

Faust (1975) from the USGS provided a study of the review and interpretation of the geologic setting of the Watchung Basalt flows in New Jersey. The key points were extracted for this section.

Location and Physiography of the Watchungs

The Watchung Mountains are in northern New Jersey in Bergen, Essex, Union, Passaic, Morris, and Somerset Counties. They form a crescent-shaped belt about 12 miles wide and approximately 40 miles long trending in a northeasterly direction from Somerville and Bound Brook in Somerset County and reaching almost to the New Jersey-New York border in Bergen County. The Watchung Basalt flows of New Jersey are exposed in three mountain ridges separated from one another by valleys eroded out of the less resistant red shales and brownish sandstones.

The basalt flows are part of the physiographic province known as the Triassic Lowland, an area bounded on the west by the Reading prong, on the north by the Manhattan prong, on the east by the Coastal Plain and the Trenton prong, and on the south by the Carlisle prong. The four "prongs" are made up of very old (probably Precambrian) metamorphic rocks, while the Coastal Plain is underlain by sediments of Cretaceous, Tertiary, and Quaternary age.

The Reading and Manhattan prongs are sometimes referred to collectively as "The Highlands" in Pennsylvania, New Jersey, and New York and that part of the Coastal Plain adjacent to the Triassic Lowlands is sometimes called the Cretaceous Overlap. The physiography of the Watchung Mountains is exceptionally well presented both textually and graphically in publications. Although the lava flows do not crop out everywhere in this belt their areal extent can be estimated to be about 500 square miles. The northern and middle part of the Watchung Basalt flows parallel the Palisade Diabase, a stratigraphically correlative intrusive unit, which crops out along the Hudson River. New York City lies abreast the northern part of the Watchung Mountains and is approximately 12 miles (19 km) distant from First Watchung Mountain. The New Germantown trap sheet, a smaller crescent-shaped flow with an inner trap knoll has been correlated with the Watchung Mountains. It lies in Hunterdon County, near the hamlet of Oldwick, N.J.

The relationship of the three mountains constitutes the Watchungs and the New Germantown trap sheet. It does not include minor fault dislocated trap masses such as the structure at Sand Brook, near Flemington, N.J. The First Watchung Mountain rises abruptly, as a rocky escarpment above the Raritan Valley and Coastal Plain which in this area is generally less than 200 feet above sea level. From Paterson to below South Orange, the escarpment ranges from 690 to 550 feet (210 to 168 m) above sea level and rises to the north to reach a maximum elevation of 885 feet (270 m) above sea level at High Mountain, a peak just north of Paterson. The height of the ridge drops from 540 to 450 feet (165 to 137 m) as it is followed farther south. The continuity of the escarpment is broken by occasional notches. The northwest slope of First Mountain is gentle and represents the regional dip of the basalts except where modified by later processes.

The valley between First and Second Mountains has an average elevation of 300 to 200 feet. The escarpment of the Second Mountain parallels that of First Mountain, and it is likewise generally continuous except for occasional notches. Its elevation in the northern part ranges from 500 to 665 feet, and in the southern part it ranges from 530 to 635 feet above sea level. The valley between Second and Third Mountain is again at an elevation of 300 to 200 feet .

The escarpment of Third Mountain is significantly different from those of First and Second Mountains in that it is discontinuous. It consists of the ridges known as Long Hill; Riker's Hill; the double curved, hook-shaped Hook Mountain and Packanack Mountain, these mountains are considered to be together with the upturned and folded edge of the sheet to the northwest of Long Hill called the New Vernon trap sheet. The valley between Third Mountain and the New Jersey Highlands is approximately 300 feet above sea level.

The apparently level crest lines of the Watchungs, parts of the Palisade Ridge, and the high hills of Staten Island led to the view that these elements define the eastward extension of the Schooley Plain (**Figure 6**).

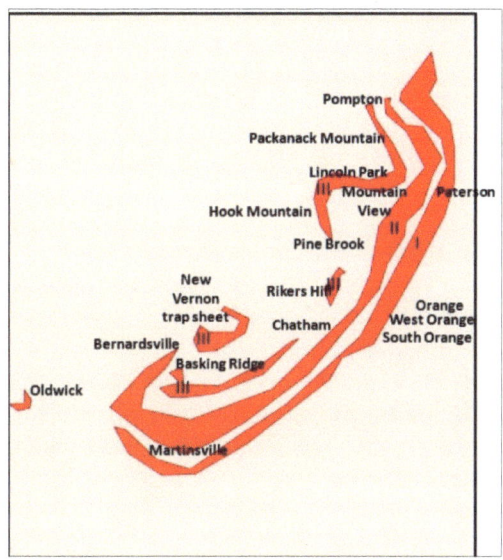

Figure 6. Sketch map showing the outline of basalt outcrops of the three Watchung Mountains. I, II, III refer to First, Second, and Third Watchung Mountains.

Geologic Setting

The geologic relations of the Watchung Basalt flows, and the geology of the Triassic Lowlands, as a unit have introduced new theories to account for the tectonic history of the region. The summary of the geology which follows is based largely on these and other sources.

At the close of the Paleozoic Era, the very large compressive forces that produced the folded Appalachian Mountains had been relaxed. At some later time, as a consequence of this change in the mechanical state of the crust, large blocks of the crust dropped down to form troughs or grabens in a region extending from Nova Scotia to North Carolina. This down faulting was accomplished by displacement of the blocks along major faults or chains of faults which made up the walls of the trough. According to some geologists, the grabens were probably not continuous from Nova Scotia to North Carolina but consisted of a series of discontinuous basins. This faulting must have occurred over a period of about 10 million years. The Triassic rocks in New Jersey all belong to the upper third of the Triassic Period, so that we can conclude that this region in New Jersey during the lower and middle thirds of Triassic time was at a sufficient elevation to be yielding sediments rather than receiving them.

The basin in New Jersey is referred to as part of the Newark basin. The geological history of the Newark basin has been interpreted in several ways. The first concept of the formation of the Newark basin was developed and summarized as follows:

1. The beginning of the basin was a down sinking of a long trough with a tendency to down warp on one side and up warp on the other. This movement was accompanied by the uplift of neighboring areas.
2. The rocks of the uplifted areas were eroded to produce the sediments, the mud and gravel, which make up the rocks formed in the trough. In New Jersey, "the basal beds disappear toward the north and that sediment came from both east and west".
3. The continued accumulation of sediments resulted in a loading over the area of deposition that produced further sinking of the trough. The greatest sinking took place along the margins where the deposition of the sediments was the greatest. The occurrence of conglomerate beds throughout the section indicated regrowth of the mountains supplying the sediments for the basin.

This necessitated a boundary consisting of a fault wall along which renewed movement took place. In New Jersey "These areas were therefore basins facing each other and bounded on their outer sides ·by faults. The western limit of the New Jersey area is still on or near this ancient boundary. The Triassic section in New Jersey contains at least 15,000 feet of sediments and igneous rocks, so that approximately "three miles of sediment was deposited in the basin in the tract of greatest subsidence. The basin floors in those places subsided at least three miles during the progress of the period, yet the sediment was sufficiently abundant to keep the basins continuously full of sediments. Slow movements intermittent in nature went forward therefore for a vast period of time, during which erosion planed even deeper into the folded and metamorphosed rocks of the rising tracts of the Appalachian system, the rock waste being swept into the intermontane basins."

4. In the Early Jurassic, extensive crustal movement took place in the Newark basin. The crust broke into great blocks by faulting and these blocks were rotated so that they now slope 15° to 20° W. In Connecticut, the Triassic rocks slope to the east. The opposite slopes of the two basins suggest that a wide mountain arch was raised between them. The raising of the arch was followed by fracture and settling along the sides. Where the principal movement of the blocks was tilting rather than elevation or subsidence relative to adjacent blocks, the upturned edge of each block would form a ridge with a steep face along the fault plane and a gentle slope following the dip of the strata. The least uplift was on the western side of the New Jersey Triassic.

In Pennsylvania, the available facts are not consistent with the hypothesis that Newark sedimentation was initiated by down faulting along the present north border or by the formation of grabens. The first sediments came from the south, where no evidence of down faulting has been found. The locations of the Newark basins, and especially the New York-Virginia one, suggest instead they were produced by deep erosion of the least resistant Paleozoic rocks, the limestones and shales of Cambrian and Ordovician age, with some down faulting along the northwest border. The more resistant crystalline rocks of the present Piedmont formed a lofty highland along the southern border of the Newark trough, while an escarpment of resistant quartzite formed much of its northern limit.

The Late Triassic tectonic history of the northeastern United States was re-examined. His field observations covered the Triassic basins from the Connecticut Valley to the Schuylkill River in Pennsylvania, and the study showed that the tectonic history of the basin is more complex than had been realized by earlier workers. On the basis of the data available, the isolated basins "represent remnants of the filling of an elongate rift valley that was 50 to 70 miles wide," and that "the stratigraphic thickness of the Triassic rocks is held to be approximately 30,000 feet."

Four discrete episodes are recognized:

1. *Initial graben subsidence, sedimentation, and basaltic igneous activity (intrusive and extrusive).*

    The basin was formed by down faulting and not down-warping. Differential relief was maintained by recurrent movement along the marginal faults. These marginal faults must have originally dipped under the graben block at approximately 70° to 75°. Now they dip 55° to 60° in many localities, but they have been rotated 15° to 20° during the longitudinal crustal arching of episode 2. Three periods of plateau basalt outpourings were associated with the initial episode of graben subsidence. Dikes and gently concave saucer-shaped intrusive sheets of similar composition to the lava flows were also emplaced during the initial episode of graben subsidence. The evidence found locally in the Palisades sheet proves that these sheets solidified before tilting of the Triassic strata. The parallelism of the Triassic strata with each other and with the three interbedded lava flow complexes, as well as the parallelism of the latter to each other and with the base of the Triassic indicates (notably in the New Jersey-New York belt) that the initial episode of graben subsidence took place without important longitudinal or transverse warping.

2. *Longitudinal crustal warping.*

    During this episode, longitudinal crustal warping in the center of the initial graben was responsible for the tilting of the New York-New Jersey-Pennsylvania and the Connecticut Valley belts so that they now dip 15° to 20° away from each other. The total uplift may have been as much as 35,000 feet.

3. *Second-generation graben subsidence and transverse crustal warping.*

   The next tectonic episode is related to the formation of the transverse folds (or warped structures, as they have been previously called) that are so prominent in the topography and geology of the Hunterdon Plateau, New Jersey. The width of the second-generation graben varies. The transverse synclinal structures show a corresponding change of size. Small synclines occur on narrow parts of the graben and large synclines occur on the wider parts of it.

4. *Fragmentation of transverse warped structures and emplacement of late intrusives (basaltic) and mineral deposits.*

5. The final episode of tectonic activity displaced the transverse folds which had been formed on the second generation graben block. In New Jersey the faults are curved and change trend from northeast southwest in the Delaware Valley to north-south in areas to the northeast.

The Flemington fault marks an important tectonic boundary in the Triassic outcrop belt. The Flemington fault shows an unknown amount of right-lateral strike slip displacement as well as more than 10,003 feet of dip slip displacement.

Stratigraphic and sedimentological studies on the Triassic basins conclude that the several troughs which make up this basin were all separate basins at the time they were in filled. The original areal extent and the distribution of the present outcrop areas of the Newark Group has given rise to two hypotheses. The local-basin hypothesis suggests that the presently existing detached basins represent essentially the original extent of the areas of deposition during Triassic time. The broad-terrane hypothesis assumes that the presently existing detached basins are only remnants of the former areas of deposition of the Newark Group, and that the missing elements have been removed by erosion. It further claims that most, if not all of the detached areas were united into a broad terrane during Triassic time. The two principal objections to the broad-terrane hypothesis with particular reference to the Connecticut and Newark basins are based on the observations that no trace of the eroded Triassic rocks between the detached basins exist and no area where eroded sediments could have been deposited was recognized.

The first objection was negated by the observations of Triassic rocks in the down faulted area of the Pomperaug valley in Connecticut, and by the recording of Triassic sediments in a well on Long Island.

The second, by the requirements of sedimentary material needed to form the coastal plain of New Jersey. The implication of the broad-terrane hypothesis suggests that the areas and volumes of Triassic basaltic volcanism were far larger than those observed today. There is general agreement amongst most investigators of the structure of the Triassic rocks in New Jersey that the Newark basin was formed as a graben. The structure was proposed to have formed by deep erosion out of Paleozoic limestones and shales. This view has not been accepted by later investigators.

The source of the sediments is attributed principally to the granitic and metamorphic rocks represented in the Trenton, Reading, and Manhattan prongs and also to the sedimentary rock of the older Paleozoic formations. The direction of transport of the sediments is still an unsettled question, and the various proposals on the subject have recently been discussed.

Structural Characteristics

All of the geologists who have investigated the volcanic rocks of the Newark basin in New Jersey agree that they are flows from fissure eruptions. The flows are described as follows:

(1) The extrusive sheets are perfectly conformable to the underlying sediments.

(2) Their upper surfaces are slag like and deeply vesiculated.

(3) The overlying sediments are totally unaffected.

(4) Trap breccias are observed at some of the basal contacts.

(5) The base of the flows are vesiculated.

(6) There is evidence of successive flows.

(7) There is a general absence of tuffs.

(8) No central cone is present.

The general absence of fragmental deposits, excepting a few local beds; the absence of craters; the large areal extent of the Watchungs; and their similarity to the great fissure eruptions of the Western United States was sufficient evidence to attribute these flows to fissure eruptions.

Each of the three Watchung Mountains was actually made up of successive flow units and was believed that the composite character of the trap sheets seems to correspond to successive flows of lava or to successive pulsations of an irregular eruption. The presence of a ropy and vesicular layer at a quarry in First Watchung Mountain indicated two successive flows were present. The study of Second Watchung Mountain, particularly in the vicinity of Bound Brook, led him to conclude that it consisted of two sheets (flow units).

In addition it was supposed without field evidence that these flow units were separated. by an inter-trappean layer of sedimentary rocks, a view denied by others, whose views on faulting were likewise unsubstantiated. The character compositing of the Third Watchung Mountain were also admittedly inconclusive.

The fact that First Watchung Mountain, in the few exposures that were examined actually consisted of two flow units-a lower flow unit and an upper flow unit separated by a thin layer of reddish argillaceous siltstone. It is important to note that this preliminary knowledge of the composite structure of the Watchung Mountains is based entirely on the structure as seen at a few widely separated outcrops. This observation was chiefly made near the nose of the syncline in the vicinity of Bound Brook while studies were chiefly in quarries near Paterson. The pillow lavas in First and Second Watchung Mountains have been described, and their outcrop has been delineated in First Mountain.

Watchung Glaciation

Northern New Jersey was subjected to several glacial invasions. Scattered deposits of pre-Wisconsin drift in New Jersey as far south as latitude 40°35'N were observed. Recent studies of this glaciation were reviewed and pointed out that the glacial deposits representing this stage lie to the west of Somerville and further that the field evidence available is so meager as to permit few additional conclusions other than its limited occurrence. No evidence of the effects of this glaciation on the Triassic sediments or lava flows has been reported. The Wisconsin Glaciation in contrast is well documented. The terminal moraine cuts across the Watchungs in a curved line from Summit to Morristown.

The Wisconsin glacier covered the northern part of the Watchungs. The thickness of the ice sheet occurred over the Triassic Lowlands. These authors estimate that the average thickness of the glacier in the area considered in this study was about one-fifth of a mile or 1,056 ft. The northern part of Third Watchung Mountain, Hook Mountain and Packanack Mountain, must have borne a considerable weight of superincumbent ice. When the glacier invaded the area, it caused a depression of the crust below it which was manifested in bending of the rock units and settling. Since the thickness of the glacier increased slowly, there was a gradually increasing pressure applied to the crust. The ice sheet maintained itself for about 2,000 years. Recent studies of the retreat of the last ice sheet in New England by radiocarbon dating, suggest that this period may be twice as long as estimates indicate.

Accordingly, the vertical component of the stress that was impressed by the weight of the ice above the rock was maintained at its maximum value for a significant length of time. The value of this stress varied from an almost negligible value at the end of the sheet to the maximum values farther north. The outcrop edge of Hook Mountain and Packanack Mountain must have been covered by at least an average thickness of the ice sheet. If the average thickness of 1,056 feet (322 m) of ice were applied and the weight of a cubic foot of ice is 57.2 pounds, (25.9 kg) we can calculate that the pressure of a column of ice 1,056 feet (322 m) high and 1 square foot would be close to 60,400 pounds per square foot (29.5 kg/cm2). These values represent the vertical normal component of the stress acting upon, and in the vicinity of the outcropping edge of the lava flow of Third Mountain.

When the glacier disappeared, either by retreating or down wasting, a reverse mechanical reaction took place, namely uplift, and the crust strove to recover from its depression. Evidence for this recovery is observed in the postglacial up warping of the shoreline of glacial Lake Passaic. Changes in elevation were calculated for the length of the lake; at Hook Mountain it amounted to about 35 feet (11m), and at the end of Packanack Mountain to about 67 feet (20 m). The evidence suggests that the largest change in altitude coincides with the greatest thickness of ice. Additional evidence of this upward movement can be seen in the glacial clays in the form of small faults with a throw of a few feet and in the recorded occurrence of occasional very minor earthquakes attributed to the adjustment of the land surface. Sheeting (sheet jointing) occurred in the basalt flows where the ice was thickest.

Stratigraphy and Geologic Age

Local Stratigraphic Relations

The Newark Group was divided into three formations made up of the Stockton Formation, which lies unconformably upon the older crystalline rocks; the Lockatong Formation, which lies over the Stockton; and the Brunswick Formation, which overlies the Lockatong and is thus the youngest member of the Newark Group. The three Watchung extrusive sheets (First, Second, and Third) were extruded near the end of Brunswick sedimentation. Each of these volcanic episodes forming the three sheets consisted of two events, and each sheet is composed of two separate flows, here designated flow units. The time lag between these multiple events was short.

The Palisade Diabase was intruded chiefly into the upper part of the Stockton Formation and the lower part of the Lockatong, but some of the Palisade Diabase intrudes the Brunswick Formation. The presence of the recurved Palisade Diabase sheet in the upper part of the Brunswick at Haverstraw, N.Y., where it reaches almost to the top ·of the series was made apparent. This observation thus dates the Palisade intrusion as near the end of Triassic sedimentation and is evidence for the conclusion that the Watchung flows and the Palisade intrusion were contemporaneous igneous events. The Palisade Diabase intrusion is also a multiple intrusive event demonstrated by its composite character on the basis of petrographic and chemical evidence.

The first intrusion to form the Palisade sill had cooled sufficiently to crystallize and to undergo some differentiation before the second intrusion was emplaced within it. The multiple events forming the sill thus show a correspondence with the events forming the sheets and their flow units. Based on the division of the Newark Group into three formations upon lithologic differences which imply diverse conditions of sedimentation, and which permit the establishment of recognizable horizons, it was observed there were no sharply marked division planes and that these formations tended to grade into one another through transitional zones, but nevertheless the ability to differentiate between them was possible. The paleontologic evidence is meager but almost all agree that the Newark Group's Upper Triassic (Keuper).

The lithologic assembly of the Stockton Formation consists of (a) coarse, more or less disintegrated arkose conglomerates; (b) yellow micaceous, feldspathic sandstones; (c) brown-red sandstones or freestones, and (d) soft red argillaceous shales. These are interbedded and many times repeated. The Lockatong Formation consists of hard dark-colored shales and dark-gray and green flagstones. It also includes carbonaceous shales, black and purplish argillites, dark-red shales, almost flagstones, and thin layers of limy shales. The Brunswick Formation consists of great thicknesses of soft red shales with some sandstone layers. Massive fanglomerate beds occur along the northwest border of the Newark Group.

The lithology of the Newark Group members have been studied by many workers. Important studies of the Lockatong Formation show that the Lockatong is a large lacustrine lens-shaped body. It was characterized by its mineralogy and interpreted chemistry. The Brunswick Formation was made up of alternations of a massive resistant layer rock (chiefly sandstone) and of less resistant beds chiefly mudstones and poorly sorted siltstones in the proportion of 43 percent of the massive resistant layers to 57 percent of the less resistant beds. The lithologic data was correlated to draw a schematic block diagram to portray the views of the stratigraphy of the Newark basin (**Figure 7**).

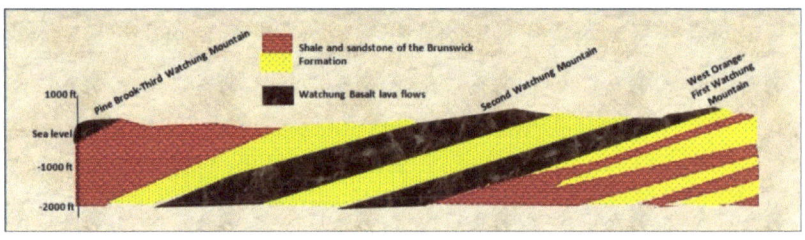

Figure 7. Generalized geologic cross section from west of Pine Brook on Third Watchung Mountain to an exposure just north of West Orange on First Watchung Mountain. Most of area is covered with glacial materials, but they generally form so thin a layer that they are not shown. The fanglomerates (conglomerates) below the lava flows do not extend for any great distance into the basin and are close to the graben wall.

The stratigraphic section of the upper part of the Brunswick Formation of the Triassic System which contains the three Watchung Basalt flows is of particular importance to this discussion; it is also shown in a geologic cross section. On the basis of the mapping that has been completed so far, each of the three Watchung Basalt flows is shown to be made up of two flow units.

This multiple character of the three Watchung Basalt flows proves that the three volcanic episodes producing them were multiple eruptive events. No inter-trappean layers were found in Third Mountain. There was a suggestion that an inter-trappean layer may occur in Second Mountain in the vicinity of Bound Brook. The flows were not found to be separated by an intervening layer, except in the northern part of Second Mountain near the Greenbrook, in North Caldwell, where a section of 6 feet (±) of sediments was measured. In First Mountain, in the Upper New Street quarry, inter-trappean materials between the upper and lower flow units were observed. This was described as a thin siltstone layer.

Although fanglomerate and conglomerate beds were generally not tied down to particular stratigraphic sections by previous workers their position with respect to the three Watchung Basalt flows in a few areas is noteworthy, and these are shown at the top of the column; just underneath the Third Watchung Basalt; and just underneath First Watchung Basalt. During the mapping of the northernmost part of Second Mountain, a thick fanglomerate bed was found at Oakland, N.J. The contact between this bed and the overlying hornfels zone was covered, but it is probably within 50 feet or less.

Fanglomerates below the flows (called conglomerates) are the most complete. These conglomerate beds were exposed south of Paterson in front of Great Notch, and in considerable force several miles north along Goffie Creek, and the conglomerates are separated from the base of the First Watchung Mountain by an inconsiderable thickness of sandstones and shales, and are succeeded eastward by fine grained materials which dip gently westward towards the coarser beds. At Paterson, a short distance east of the conglomeratic outcrop in the gorge of the Passaic, a well was bored 2,100 feet through fine grained sandstones and shales. The significance of the fanglomerates as marking the intermittent re-growth of mountains whose perennial waste kept supplying material for the deposits of the basin was mentioned.

The relations of the fanglomerates to the lava flows and the presence of a layer of sandstone and shale separating the conglomerate from the base of the flow, a conclusion that the volcanic episode forming the lavas of First Watchung Mountain took place shortly after the faulting along the boundary wall occurred. Similarly a fanglomerate bed occurs about 50 ft (15 m) (±) below the lower flow unit of Second Watchung Mountain. Fanglomerate beds also occur just under the base of the flow forming Third Watchung Mountain.

All of these stratigraphic observations attest to the relation that faulting along the boundary wall was followed by the formation of a fanglomerate and that shortly thereafter the lava poured out. These relations of the fanglomerates to the flows show clearly that the periods of faulting along the graben walls were unrelated to the volcanism.

Finally, there exists a dilemma over the thickness of the Newark Group. The thickness was originally calculated to be 27,000 feet (8,230 m); but then, because of suspected hidden faults that might be present to increase the true thickness, the thickness was reduced down to 15,000 feet (4,570 m). A very few faults have been found but they are only a few feet in extent. Taking these data for a basis of calculation it would make a formation not less than 25,000 feet (7,620 m) in thickness. The problem was reexamined by others and calculated 25,000 feet (7,620 m) with the reservation that significant hidden faults were not present. After intensive field study, it was calculated the thickness of the Stockton Formation was estimated to be 4,700 feet (1,433 m); of the Lockatong Formation it was estimated to be 3,600 feet (1,097 m), and of the Brunswick Formation to be 12,000 feet (3,658 m) giving a total thickness for the Newark Group of 20,300 feet (6,187 m).

Two faults were examined in the basalt of the First Watchung Mountain with throws of 75 (22.9) and 8 feet (2.4 m) respectively, with awareness of earlier suspicions of hidden faults. Undiscovered in field studies, the previous estimate was reduced to two revised estimates. The first reduction was attributed to the Stockton having a thickness of 2,300 feet (701 m), the Lockatong of 3,500 feet (1,067 m), and the Brunswick of 6,000 feet (1,829 m) making a total thickness of 11,800 feet (3,597 m). The second estimate made the Stockton 3,100 feet (945 m), the Lockatong 3,600 feet (1,097 m), and the Brunswick 8,000 feet (2,438 m) with a total thickness of 14,700 feet (4,481 m). The exposed Triassic section along the Delaware River had a thickness of 12,000 feet (3,658 m). The Triassic of the northeastern United States, and for New Jersey was restudied for a calculated thickness of 30,000 feet (9,144 m). Estimates of 16,000 to 20,000 feet (4,877-6,096 m) were arrived at after a detailed study of the Lockatong Formation and its relation to the Stockton and Brunswick Formations.

The evidence of the recent studies cited above may resolve the dilemma, because it supports the greater thickness calculated by early workers. The often cited hidden faults have not been found and the lower estimates based on this assumption of their existence are not supported by field evidence in that part of the Newark basin containing the Watchung flows and the Palisade sill.

Two thicknesses, of 25,000 and 11,800 feet (7,620 and 3,597 m) respectively, for the Newark Group sediments are used here in the calculations because of the uncertainty as to the true thickness of the group.

Upper Triassic Fissure Eruptions in New Jersey

The timing of the fissure eruptions in the Newark Group may be estimated on the basis of some recent summaries of the paleontological evidence; studies indicate that the Newark Group is Late Triassic in age. Because the Triassic Period was 45 million years long and the Newark Group was deposited in less than one-third of that time that the deposition took place in a time span of at least 10 m.y.

Using an estimate of the time span of deposition, the thickness of the strata above and below the flows, a provisional estimate of 300 feet (91 m) of sediments above Third Watchung flow, and assuming a total thickness of the Newark Group of 11,800 feet (3,597 m), then it may be calculated that at least 80 (79.7) percent of the sediments were deposited before the First Watchung flow appeared. If the greater thickness of 25,000 feet (7,620 m) is assigned to the Newark Group, then at least 90 percent of the total thickness of the sediments was deposited before the flows appeared. These calculations suggest that the outpouring of the lava from the fissures started at 2.5 or 1 m.y. before the close of the Triassic Period.

It is thus apparent that the deep fissures from which the lavas were erupted were developed very late in the Late Triassic Period. They were associated with the initial episode of graben subsidence. Using an average figure of 600 feet of sediments between First Mountain flow and Second Mountain flow and a total thickness of 11,800 and 25,000 feet (3,597 and 7,620 m), a corresponding time interval of about 600,000 or 240,000 years elapsed between these outpourings of the lavas. The thickness of sediments between the Second. and Third Watchung flows is approximately 1,500 feet (457 m), and this value yields a corresponding time interval of 1.5 million or 600,000 years between the flows. The calculated time intervals support the belief that fissuring in the graben was intermittent.

These calculations assume that the overall rate of deposition of the sediments was, on the average, relatively constant. This assumption implies a more or less uniform rate of supply of sediments and some form of tectonism that will periodically down fault the basin with respect to the source area.

The different lithologic types of the Connecticut Triassic are due to variations of one fundamental sedimentary process: rapid erosion and deposition under savanna climate, and that this geologic setting assured an abundant supply of sediments at an almost periodic rate of supply. The almost periodic reappearance of the fanglomerates along the fault walls is suggestive of at least some periodicity to the character of the sedimentation between successive periods of faulting. The principal assumption is open to criticism by some stratigraphers, the use of rates of sedimentation for the rocks in the British Isles and observations that rates of sedimentation are very elusive, however, and they can only be deduced at all reliably either on a very small scale or a very large one.

Average rates of deposition were used as a method to dispute the calculations of Lord Kelvin on the age of the earth, a good example of the use of the assumption on a tremendous time scale. Rates of sedimentation were studied in the Upper Carboniferous rocks of Great Britain, and they emphasize two aspects in the concept of rates of sedimentation, an overall rate, and a specific rate. The overall rate of sedimentation is the mean rate of sedimentation during a given unit of time in which phases of sedimentation may be interspersed with periods of non-deposition or erosion (sedimentation remaining dominant). The specific rate of sedimentation is the rate of sedimentation at any one moment of time.

The specific rate is very difficult or well nigh impossible to determine in most studies. The overall rate of sedimentation is the concept used in these calculations. The basis of stratigraphic and other measurements in sections exposed along the Delaware River in New Jersey and Pennsylvania, the rate of initial deposition of the Stockton, Lockatong, and Brunswick Formations were estimated. The thicknesses are 6,000, 3,772, and 6,232 feet (1,830, 1,150, and 1,900 m), respectively, with a total thickness of 16,006 feet (4,879 m). The fastest rate of initial deposition is 600 mm/1,000 years (= 1.97 ft/1,000 yr) for the Stockton, the slowest is 215 mm/1,000 years (= 0.705 ft/1,000 yr) for the Lockatong, and the rate for the Brunswick Formation is intermediate, 305 + mm/1,000 years (1 ft/1,000 yr). The uniform rate of basin was estimated to be sinking about 215 mm/1,000 years (=0.705 ft/1,000 yr) during the entire Newark time, which is the same as the rate of deposition for the Lockatong Formation. From these data he estimates the total Newark episode to have been 22.75 m.y. in duration.

Several other possible interpretations of the data were made, arriving at an estimate for the minimum of 14.75 m.y. and maximum of 22.75 m.y. for the duration of the Newark episode and for a minimum of 215 mm/1,000 years and maximum of 325 mm/1,000 years for the average rate of basin sinking and sediment accumulation.

If these data are used and calculated as before, the following values are arrived at for 16,000 feet (4,877 m) total thickness-85 percent of the sediments were deposited before the First Mountain flow appeared; 3.75 percent of the sediments were deposited after First Mountain flow solidified and before Second Mountain flow appeared, 9.375 percent of sediments were deposited after Second Mountain flow solidified, and before Third Mountain flow appeared; 1.875 percent of the sediments were deposited after Third Mountain flow solidified. Using the lower time duration of 14.75 m.y., this yields 12.54 m.y. before the First Watchung flow appeared, 0.55 m.y. between First Watchung flow and Second Watchung flow, 1.38 m.y. between Second Watchung flow and Third Watchung flow, and 0.28 m.y. before the close of Triassic deposition.

If the longer time span of 22.75 m.y. is used, the figures become 19.34 m.y., 0.85 m.y., 2.13 m.y., and 0.43 m.y. respectively. The time intervals between the appearance of First and Second Watchung flows and between Second and Third Watchung flows as calculated from the time interval of 14.75 m.y. are closely similar to those just calculated for a section of 11,800 feet (3,597 m) of sediments.

Igneous Activity Geologic Age

The earliest work on the dating of the igneous activity during the Triassic in eastern North America used the "helium method" for age determination and dated a specimen of basalt taken from the base of the First Watchung Mountain flow in a quarry on Valley Road, a quarter of a mile north of Notch Road, Great Notch, Clifton, N.J. to be 180 ± 11 million years old. Three diabase specimens were dated taken from the Palisade intrusion at Kings Bluff, Weehawken, N.J., at a distance of 5 to 10 inches (13-25 cm) above the basal contact with the Triassic sediments; 40 inches (1 m) above the basal contact; and 40 feet (12m) above the basal contact to be 155 ± 8, 165 ± 9, and 165 ± 8, m.y. old. For the diabase of the Connecticut Valley it was found that two specimens of the West Rock sill in a quarry at New Haven, Conn. gave ages of 170 ± 12 m.y. and a dike in the same quarry had an age of 175 ± 9 m.y. A drill-core specimen of basalt from 464 feet (141 m) below the surface of the Cape Spencer Flow, Nova Scotia, Canada, yielded an age of 160 ± 8 m.y.

The "helium method" is very sensitive to the loss of helium by the host, and as a result it tends to give ages which are younger than they should be.

Using the K-Ar method, a specimen of columnar basalt from First Watchung Mountain was dated and found that the apparent age was 79 m.y. For the Palisade Diabase sill, biotite was dated separately from the upper fine zone of the sill and it was discovered that an average age of 190 ± 5 m.y was measured. The low value for the age of the Watchung Basalt was attributed to the poor retentivity of the argon by the basalt. The problem of dating Triassic mafic igneous rocks was re-examined, and they were analyzed by the K-Ar method from the same sample of biotite from the Palisade Diabase. An age of 192 m.y. was measured as compared against the values of 190 and 196 m.y. measured previously. The results agree within 1 percent of the Lamont-Doherty date. As a result of the study of the Upper Triassic Newark Group, however, a conclusion of the dates for the Palisades diabase, although reproducible and internally consistent are all suspect.

The most recent evaluation of the age data of the Triassic igneous rocks of New Jersey is that the K-Ar dating of the biotite from the Palisade sill was determined to be a "critical point" on the Phanerozoic time-scale with the values of 186 ± 5, and 194 ± 5 m.y. The observation resulted that In any case the stratigraphical assignment contained an element of uncertainty, that is, the correlation of the sill with the Watchung basalt was uncertain.

The paleomagnetic study in which the three Watchung flows of First, Second, and Third Mountains were correlated with the Holyoke Basalt flow of Massachusetts is difficult to reconcile with the stratigraphy. This correlation means that the extrusion of the Holyoke Basalt flow, 300 to 600 feet (91-183 m) thick in Connecticut and 575 feet (175 m) thick in Massachusetts could have occupied as much time as the following events in New Jersey:

1. Outpouring of approximately 600 feet (183 m) of lava to form the First Watchung flow.
2. Deposition of approximately 600 feet (183 m) of sediments.
3. Outpouring of approximately 750 feet (229 m) of lava to form the Second Watchung flow.
4. Deposition of approximately 1,500 feet (457 m) of sediments.
5. Outpouring of approximately 300 feet (91 m) of lava to form the Third Watchung flow.

This time interval in New Jersey spanned at least 1 to 3 m.y. Such a correlation would suggest that during this time the area of the Holyoke sheet was high land and that there were multiple extrusions to form the sheet. Evidence for such a high land in Connecticut would be extensive erosion within this compound sheet or at its upper surface.

Regional Stratigraphic Relations

A comparison of some-time intervals between Triassic events in New Jersey and in Connecticut is given here because of its relevance to the problem of assessing the state of stress in the graben. The rate of relative depression of the graben blocks during Triassic time cannot be accurately calculated. If it is assumed that the relative downward movement was uniformly continuous, the rate of depression for a section of 10,000 feet (3,048 m) of sediments in 10 m.y. would be approximately a foot (0.3 m) of depression in a thousand years. For a section 25,000 feet (7,620 m) thick, it would be a foot (0.3 m) in 400 years. A value of 0.7 foot (0.2 m) per 1,000 years was calculated based on the duration of Late Triassic time of 15 m.y. and assuming a uniform rate of basin sinking. The repetition in the geological record of fanglomerates (conglomerates) at the graben walls against the horst, however, suggests that the relative depression was discontinuous--neither uniform nor a succession of a few major movements--but rather spasmodic in response to the unbalance produced by the weathering of the horst and the removal and distribution of the detritus into the basin. Such relations between time and events may be described as almost periodic.

The best exposed and studied sections of fanglomerates and conglomerates are those in southern Connecticut measured. Fanglomerates were found to have a maximum thickness of 200 to 300 feet (61-91 m) and that they never extended more than about half a mile from the fault plane. The position of the fanglomerates with respect to the lava flows were explained to be as follows: The fanglomerate is found *below* the lower basalt flow, immediately *below* the main sheet, *below* the upper basalt flow, and *above* all the flows. If the appearance of a fanglomerate along the fault wall marked the early stage of the weathering cycle in the topographically higher granitic horst blocks, then it can be equated that the number of fanglomerates observed in sections with the number of episodes of faulting; for the reappearance of fanglomerates means the end of a former period of faulting. From the section given, four periods of intermittent faulting bounding the three lava flows are recognized.

Combining the observations with the stratigraphic section of the Newark Group in Connecticut, the observation that the lower basalt flow erupted shortly after a fanglomerate appeared can be made; there then followed a deposition of nearly 1,000 feet (305 m) of sediments, and near the end at the top of the section a fanglomerate appeared; this event was followed by the outpouring of the thick main sheet of basalt and on it were deposited nearly 1,100 feet (335 m) of sediments with fanglomerates near the top; thereafter the upper basalt flow appeared, and this was followed by the deposition of about 2,000 feet (610 m) of the lower zone of the Portland Arkose, which contained at least one fanglomerate above the lava flow.

It is thus apparent that the lavas did not appear at the time a given period of faulting ended but only after the time interval necessary to allow the disintegration, weathering, and transportation of part of the granite rocks of the horst to form a fanglomerate at the graben wall. If it is assumed for purposes of estimation that the values calculated for the rate of deposition in the Newark basin of 1 foot (0.3 m) of sediment accumulating in a thousand years for the section 10,000 feet (3,048 m) thick and 1 foot (0.3 m) in 400 years for a section 25,000 feet (7,620 m) thick, this would indicate that approximately a million years or alternately 400,000 years elapsed between the eruptions of the lower basalt flow and the main sheet of basalt and similarly about 1.1 m.y. or alternately 440,000 years lapsed between the main sheet of basalt and before the upper basalt flow was erupted. These time-interval values are of the same order of magnitude as those for the stratigraphic section in New Jersey where the corresponding values are 600,000 or 240,000 between First and Second Watchung, and 1.5 m.y. or 600,000 years between Second and Third Watchung. The good correspondence between the times elapsed between eruptive events in New Jersey and Connecticut casts doubts on the paleomagnetic correlations.

The Pomperaug valley near Southbury, Conn., is a small down faulted part of the former Triassic section. It is isolated from the principal outcrop area of the Triassic in Connecticut and is 15 miles (24 km) west of it. It contains the lower part of the section and includes two of the lava flows. The South Britain Conglomerate, which underlies one of these lava flows was mapped. It is thus evident from field relations in New Jersey, Connecticut, and Massachusetts that the stratigraphic position of the fanglomerates (conglomerates) below the lava flows is significantly consistent. This observation suggests that the extrusions of lava did not coincide with the period of faulting but rather were later than these tectonic events.

An estimate of the time between successive down faultings along the wall of the fault system is also suggested by these calculations involving the eruptive events. They suggest that the time interval between the fanglomerates, for the upper part of the Triassic section in southern Connecticut, was of the order of half a million years and that an episode of faulting likewise was regenerated on the order of half a million years with the production of an uplifted horst block.

Evidence for this short duration of the periods between faulting is also supplied in a study of the sediments, concluding that the Connecticut Triassic and the different lithologic types of rocks are due to variations of one fundamental sedimentary process: a rapid erosion and deposition under a savanna climate. As an index of the rapidity of erosion, transportation, and deposition, the observation that the feldspars are quite stable was made. The further observation was made that the mafic minerals decomposed to form iron oxides and not a montmorillonitic type of clay. Concerning the feldspars, they are generally fresh, but that all degrees of weathering and decomposition are present and that most of the alteration is primary-some is post-depositional.

The feldspars make up on the average 30 percent of the Triassic sediments of Connecticut, and the fragments are generally angular or sub angular in shape. This persistence of the feldspars is significant as an index of the shortness of the time span that these grains could have been exposed to weathering. The feldspars are relatively easily attacked under weathering conditions at pH values in the range of 3-11 and that at the higher temperatures corresponding to tropical conditions the feldspars are dissolved in solutions that are even in the neutral range of 6-7 pH units. Although it is not possible at present to quantitatively correlate persistence of the feldspars in sediments with their chemical resistance in a geologic environment, their presence in large amounts in a relatively stable state does support the opinion that the products of weathering did not remain for great lengths of time on their source rocks. The oft-repeated appearance of large quantities of feldspar grains throughout the stratigraphic section attests to their renewed production from freshly weathered granitic rocks. Such fresh rocks would be made available by the postulated intermittent, that is, non-uniform and not quite periodic movements along the fault walls during the process of sedimentation.

Sites of the Fissures

The sites of the fissures from which the lavas flowed into the Triassic basin have not been located. No evidence for their existence, such as lava filled feeder dikes has been found in the underlying Triassic sediments well exposed in outcrop to the southeast. The possibility that they exist here and are covered by soil must be considered. The observed greater stability of the basalt over that of the sediments in the Watchungs proper would suggest that the dikes should form topographic "highs." No trace of such dikes has been observed in the sediments of the valleys between the flows. The supposed stability of the trap dikes has been questioned, noting that under the climatic conditions obtaining in Virginia and North Carolina the dikes were frequently deeply decayed, sometimes to depths of 50 feet (15 m), and that they did not form ridges nor otherwise manifest their presence in the topography.

Intrusive dikes have been found in the vicinity of the Palisade Diabase sill. The dikes are extensions of the sill itself. The difficulty of finding concealed fissures under the flows has been emphasized in accordance with the following: the chief difficulties to consider are first the thickness of the flows themselves, which blanket the surface and conceal all that is below; and secondly, the low probability of finding a feeder dike, of the order of 5-feet (1.5 m) width, exposed by erosion or topographically expressed in a depression, or encountered in excavations or drillings.

A filled fissure, now a dike, has been found cutting through the basal part of the lowermost flow in First Watchung Mountain near West Orange, N.J., in the cut for U.S. Highway 280. This fissure furnished lava to the still liquid part of the lowermost flow. Evidence for a probable dike was also observed at Moggy Hollow, at the southern part of Second Watchung Mountain. The diabase dikes which have been found in the crystalline rock along the border fault system have been suggested as possible feeder dikes. Unless, however, field evidence demonstrates an unequivocal relation between the flows and the supposed feeder dikes, the assignment of a Triassic age to these dikes is unwarranted. Dike swarms in the Appalachians using paleomagnetic methods show that most of the extensive dike swarms cutting Triassic and older formations probably intruded in a time of regional tectonic and magnetic activity, distinct from the late Triassic tectonogenesis. A Jurassic age for the dikes is suggested.

Relations of these dikes were studied showing that (1) most dikes locally have a common trend; (2) that these local common trends vary systematically from one part of the Appalachian region to another; and (3) that the trends of the dikes are everywhere discordant to the trends of the structures of the enclosing rocks, and they are everywhere straighter than these structures. Trends of these dikes in eastern North America, West Africa, and northeastern South America were studied. The trajectories of the principal stress indicated by the patterns of the dikes consist of ages to be Late Triassic to Early Jurassic. The conclusion to be drawn from these studies of the diabase dikes is that they are unrelated to the principal Triassic volcanism.

Pre-eruptive Character of the Graben Floor

The nature of the sedimentary floor onto which the lavas flowed may be deduced from a study of the contact relations. The conformable nature of the bottom of the flows with the underlying sediments as seen in abundant outcrops on the scarp slope of Third and Second Mountains shows that the fissure eruptions took place in open country, a fact deduced independently from stratigraphic studies of the non-marine sediments that contain the flows. The natural remnant magnetization in both the sediments and the Watchung flows in the Newark Group sediments demonstrate an inclination error of paleomagnetic significance is not present. In other words, the sediments were horizontal, or nearly so, when the flows poured out over them. In general, along the scarp slope of Third Mountain the lowermost part of the lower flow unit overlying the Newark sedimentary rocks is heavily vesiculated for a distance of about 1 foot (0.3 m) above the contact, and there after the vesiculation decreases rapidly as one ascends along the scarp slope. Locally, as at the quarry at Millington the vesiculation in the lower flow unit overlying the sediments is in a zone of about 20 feet (6 m), but the abundance of vesicles varies from the top to the bottom of the zone and is heaviest closer to the contact. The actual contact between the lower flow unit and the hornfel sized sedimentary rock is sharp.

The features recognizable in the contacts exposed on Third Mountain suggest that the flows poured out during a relatively dry season when the sediments were probably just barely moist or, in places, nearly dry. They further suggest that the flow of lava was laminar and that there was a constant and continuous supply of hot fluid lava. The contact of the upper flow unit with the lower flow unit of the basalt in Third Mountain wherever it has been examined is always sharp and shows no evidence of inter-trappean sediments.

The base of the upper flow unit is generally vesicular for a short distance of several feet (1 m ±) or so above the contact. At the quarry in Millington the pahoehoe surface of the top of the lower flow unit is visible, and the total thickness of the vesicular zone consisting of the vesicular top of the lower flow unit, including the vesicular base of the upper flow unit is possibly 15 feet (4.6 m) with varying degrees of vesiculation.

The contact relations between the vesicular base of the lower flow unit of Second Mountain and the sedimentary floor of the graben is variable. Over long stretches it is similar to that of Third Mountain, but at a number of places there is a slight development of pillows as in the cut for U.S. Highway 78 near the village of Pluckemin. Elsewhere, near the temporary ends of a flow, or a former surge of lava, tongues of lava may form which have a resemblance in shape to toes. In some areas both pillows and toes may form suggesting that the lava was flowing over an area with braided drainage. The appearance of the pillows and toes suggests that the flow of lava may have been temporarily halted and later resumed after overcoming a temporary obstacle.

The contact between the upper flow unit and the lower flow unit is sharp, generally lightly vesicular, but in a few areas thin zones of pillows or pahoehoe toes are observed. The presence of an inter-trappean layer of Brunswick sediments near the Greenbrook in North Caldwell was observed. These Newark sediments are about 6 feet (1.8 m) thick. The lower contact of the upper flow unit here is vesicular for a distance of about 1 foot (0.3 m) or so. Only a small number of outcrops in First Mountain were examined, but at Montclair, South Orange, and at several localities in Paterson and elsewhere the lower flow unit is thinly vesicular and rests on hornfel sized sediments. The contact between the base of the upper flow unit and the vesiculated top of the lower flow unit is separated at several localities in the vicinity of Paterson by a thin inter-trappean layer of argillaceous siltstone.

The base of the lava flow above this siltstone is composed of pillow lavas. Measured thicknesses of 50 to 75 feet (15-23 m) were made for the pillow zone at the Upper New Street quarry along with further examined localities in the vicinity of Paterson. The occurrence of pillows was used as indication of the former occurrence of a playa lake; the extent of this lake which he called Lake Paterson was mapped.

Surface Character of the Vesicular Tops

The study of the Newark rocks was an opportunity to see many small farmers quarries which were opened for road material, and exposed many shallow road cuts. The farmers quarries were commonly dug out of the upper vesicular zone of the upper flow units because it was the easiest basalt to quarry. The exposures provided the chance to observe the pahoehoe tops of the flows and they were described as follows: Upper contacts have not been observed in many cases, but the upper surface of these sheets is frequently vesicular, amygdaloidal, and scoriaceous. Locally, a thin layer of water worn trap particles intermixed with red mud occurs between the vesicular trap and the unaltered typical red shales, or the vesicles are filled with red mud. The overlying shales conform to the slightly irregular surface of the trap. In frequent exposures, the rolling-flow structure named by the Hawaiian Islanders as pahoehoe is visible.

Nowhere have any tongues of lava been found extending from the main sheet into the neighboring shales. A few good contacts of the Newark sediments overlying the vesicular top of the flow were observed, and these showed the pahoehoe surface and the conforming sediments. Such contacts can best be seen under red shale knolls.

Areal Extent and Volume Relations of the Watchung Basalt Flows

The areal extent and the volume relations of the Watchung Basalt flows are not generally appreciated. Outcrop areas and drilling operations for water supplies show the area to be at least 500 square miles (1,295 km2). Using an average thickness of 600 feet (183 m) for First Mountain, this amounts to a volume of 56.8 cubic ·miles (237 km3); an average thickness of 750 feet (229m) for Second Mountain, this amounts to a volume of 71.0 cubic miles (296 km3); and an average thickness of 300 feet (91 m) for Third Mountain, this amounts to a volume of 28.4 cubic miles (118 km3). The total volume for the three volcanic episodes is approximately 156 cubic miles (650 km3) as currently delineated.

According to views on the former probable width of the graben in New Jersey, this volume may amount to only a vestige of their former volume. The width of the graben in New Jersey is estimated to be 50 to 70 miles (80.4-112.6 km).

Using a width of 50 miles (80.4 km) and assuming for purposes of calculation a length equal to the present eroded remnant of 40 miles (64.4 km), these values for the three Watchung Mountains would become 227 cubic miles, 284 cubic miles, and 114 cubic miles (946, 1,184, and 475 km3) respectively with a total volume of 625 cubic miles (2.6 x 103 km3).

The igneous rocks of the eastern Triassic basin extend from Nova Scotia to South Carolina over a distance of about 1,200 miles (1,931 km). The areal extent and volume relations for the lava flows in the Triassic basins including Connecticut, Massachusetts, New Jersey, and Pennsylvania together with the extensive flows in Nova Scotia and the lesser volcanism in Maryland, Virginia, and the Carolinas suggests a rather considerable former areal spread and a volume of possibly as much as 3 x 103 cubic miles (1 x 10·1 km3).

Their extrusion in the Late Triassic Period constituted a major geologic event in eastern North America. Compared to the lava volumes of the fissure eruptions of the Deccan in India, the Columbia and Snake River areas in Western United States, the Brito-Arctic region (Thulean province), and the Parana basin of Paraguay, Uruguay, and Brazil, the lava flows of the Newark basin were smaller. The volume of the Deccan flows is 1.24×10^5 to 25×10^5 cubic miles (5.18×10^5 to 10.4×10^6 km3); for the Columbia River Plateau as 0.5×10^5 cubic miles (2×10^5 km3) and for the Lake Superior region, conservative estimate) as 60,000 cubic miles (2.5×10^n km3).

Geologic Processes Responsible for the Jointing Systems

Joint systems were developed in the Watchung Basalt as a result of stress fields set up by three geologic processes which acted upon the rocks of the graben. Cooling of the lava flows gave rise to the cooling joints. The tectonic forces associated with the post-Triassic folding and the accompanying deformation of the graben produced the tectonic joints. During the Pleistocene Epoch, the northern part of the eroded remnant of the former graben was covered by a continental glacier which compressed the rocks beneath it. After the ice melted, the resulting decompression of the basalts is thought to have produced sheeting.

Petrochemistry

The Watchung Basalt flows and the Palisade Diabase intrusion were probably derived from the same magma chamber. Accordingly, the chemistry of the Watchung Basalt is of prime interest in discussing the course of differentiation in the Palisade Diabase sill, but surprisingly few analyses of the basalt exist. Eight "superior" chemical analyses were used to interpret the chemistry of these rocks. Five of the analyses were for First Mountain and three for Third Mountain. These samples were placed in their approximate stratigraphic position-designated as lower, middle, and upper layers-in order to discuss their significance. The analyses of the rocks from First Mountain suggested that Al_2O_3 and (Fe_2O_3+FeO) remained essentially constant throughout the thickness of the sheet and that Na_2O is at a minimum at the middle layer, whereas CaO is at a maximum. Silica increased only slowly towards the top of the sheet, and conversely MgO decreased toward the top.

The rocks from Third Mountain showed some differences from those from First Mountain in that there was a noteworthy lower amount of SiO_2; that MgO, Al_2O_3, and CaO were also lower, but less so; and that $(Fe_2O_a + FeO)$ was noticeably higher. The Na_2O and TiO_2 were also higher in the rocks from Third Mountain. This suggested that the striking differences again appear in the MgO, CaO, Na_2O and K_2O, particularly in the higher CaO and the lower MgO and Na_2O in the second layer.

The specimens from Third Mountain, that were analyzed, came from the quarry at Millington, N.J. When this quarry was examined only the upper flow unit as it was characterized was exposed, so these analyses are from that flow unit. The specimen from O'Rourke's quarry, West Orange, N.J., comes from the base of the lower flow unit of First Mountain. Four analyses on two pillow lavas from the Upper New Street quarry (Burgers quarry) and the Lower New Street quarry, in Paterson, N.J. were analyzed. Both the glassy-looking crusts and the finely crystalline basalt inside of the pillows were analyzed and demonstrated that the glassy-looking crust is not the original glassy selvage and therefore that the analysis of this crust does not define the original lava chemistry. According to this interpretation these specimens came from the base of the upper flow unit of First Mountain.

The relation of the Watchung Basalt to the so called Plateau Basalts, which was correctly related petrogenetically to the basalts of the Thulean province, the Deccan, the Columbia River plateau, and elsewhere. The chemical similarity of the flood basalts resulted from fissure eruptions and attributed their fluidity to their notable content of ferrous oxide.

A comparison of the average of 36 analyses of specimens from the Millington quarry analyzed for this study shows reasonable agreement for SiO_2, Al_2O_3, ($FeO+Fe_2O_3$), MgO, TiO_2, and P_2O_5, with small differences for CaO, Na_2O, K_2O, and MnO. Studies of the section at Millington suggest that these slight differences are unimportant and that they probably arise in part because two of these rocks may actually have been slightly altered. A carefully selected set of 22 analyses of rocks, judged to be the least altered, from within the larger set of 36 analyses was averaged. This average is tentatively considered to be representative of the upper flow unit at the Millington quarry.

The average sample for the Watchungs of eight analyses is a very small sample, but since it is derived from the average of five analyses from First Mountain basalts and three from Third Mountain basalts it is more representative than the single set of analyses from the Millington quarry. A comparison of average with the averages of the Whin sill (5 analyses), normal tholeiitic basalt, Plateau basalt, and the Columbia River basalt (13 analyses) shows its close affinity to tholeiites. The conclusion that the Watchung Basalt is tholeiitic was reached. The Thornton-Tuttle differentiation index and the calculated normative plagioclase show the variation within those averages. The basalts all carry quartz in the norm.

The close chemical resemblance between the rocks of the Whin sill and the basalts, dikes, and diabase of New Jersey and Connecticut were noted. A comparison of the average of 22 selected analyses of basalts from the upper flow unit of Third Watchung Mountain with the average Palisade basal-chilled dolerite shows some important differences. The Watchung Basalt has lower SiO_2, Al_2O_3, MgO, Na_2O, K_2O, and MnO than the basal-chilled dolerite and about the same amount of CaO, TiO_2, P_2O_5, and MnO. The basalt is significantly higher in total iron oxides ($FeO + Fe_2O_3$). These differences are reflected in the norm by the appearance of more quartz, albite, and magnetite; slightly less diopside, and less orthoclase, anorthite, and much less hypersthene than in the basal-chilled dolerite. The Thornton-Tuttle differentiation index of the basal-chilled dolerite is lower than that of any of the tholeiites, and the plagioclase in the norm is much more anorthitic.

Megascopic Characteristics of the Basalt

The basalts of the upper flow unit of Third Mountain are uniformly fine grained, with an average grain size of less than 1 mm. Small lenses, composed almost entirely of crystals 2--3 mm across, appear rarely in an excavation near Myersville and at the cut near Pine Brook. These lenses occur in the section about two-thirds of the distance up from the base of the flow unit. They represent the last liquid to solidify, and because of the increased volatile content of this last liquid, they have completely crystallized. The lower flow unit of Third Mountain is also fine grained, but small phenocrysts about 2 to 3 mm across of feldspar and pyroxenes are sparseiy distributed in the rock.

The basalts of Second Mountain show much more variation in grain size. In the lower flow unit they are dense and fine grained near the base, but they coarsen slowly as one ascends in the section until at a zone about two-thirds of the way up from the base, the rock becomes coarse grained and resembles a gabbro. The grain size diminishes rapidly above this zone and becomes fine grained to the top. The upper flow unit is more uniform than the lower flow unit, and it has pockets of coarsely crystallized rock above the middle of the unit. These crystals are, however, not as large as those in the lower flow unit. This coarse facies does not form a continuous zone but consists only of a series of discontinuous lenses at about the same horizon. The zones of pillows at the base of the flow units appear to ·have their individual pillows coated with a glassy-looking rind. These rinds are now however altered, and the rind may crumble when handled. The basalts of First Mountain appear to be fine grained.

The color of the basalt is normally black, frequently with a greenish tint, but grayish zones are also common. There is considerable variation in color in the altered zones. On weathered surfaces, the basalts commonly have a yellow coating of iron oxides. The color varies from grayish-orange (rock color chart National Research Council 10 YR 7/4) to moderate yellowish brown (10 YR 5/4). On quarry walls which have been undisturbed for long periods of time, they may have a reddish brown (pale brown, 5 YR 5/2) coating formerly referred to as a patina when quarries were worked for dimension stone. In hydrothermally altered zones, they may be greenish owing to considerable quantities of chlorite.

The tops and bottoms of the flow units are vesiculated. The vesiculation is generally light at the basal contact of the lower flow unit overlying the sedimentary rocks. The size and the frequency of the bubbles increase as the contact is approached. At a distance of 6 inches to 1 foot (0.15-0.30 m) above the contact the vesicles are commonly the size of a pinpoint and then increase to as much as one-fourth of an inch (0.6 cm) in diameter at or near the contact. Less frequently, the individual bubbles are larger in diameter. Locally, the zone of vesiculation may be thicker.

The top of the flow unit may be as much as 10 feet (3.0 m) thick, and it is heavily vesicular. In this zone the larger vesicles are at the top. Locally, the concentration of larger bubbles may be so great that clustering produces a scoriaceous variety of basalt characterized by coarse bubbles forming a cellular structure. Bubbles are scattered sparsely throughout the flow units, and sometimes they collect into single wavy layers, thin zones about one-fourth inch (0.6 cm) thick, and much more rarely into zones 1 foot (30 cm) or so thick. Such zones of vesicles have been observed near the top of the flow unit but well below the vesicular top.

The top of the upper flow unit has a vesiculation pattern similar to that of the top of the lower flow unit, but the basal part of the upper flow unit may be more heavily and more thickly vesicular than the corresponding base of the lower flow unit. The actual tops of the two flow units have the pahoehoe form of a billowy undulatory surface which in cross section resembles a sine wave.

Pipe vesicles, tube like cavities as much as 4 or 5 inches (10 or 13 cm) long and oriented with their largest axis (the tube axis) normal to the contact are most commonly observed at the base of the upper flow unit where it lies in contact with the top of the lower flow unit. In the Millington quarry, where the contact could be seen, the base of the upper flow unit conformed to the undulatory top of the lower flow unit, and the orientation of axes of the pipe vesicles also reflected the curvature.

Field observations of the pipe vesicles, also called pipe amygdules, at the base of the lava flow where it rests on an underlying flow or on the Triassic sediments of the Brunswick Formation are in accord with those in South Africa. In an extensive field study in South Africa and elsewhere, an observed pipe vesicle (amygdule) in the upper or middle part of the flows was never observed.

This view of their origin was that the gases which formed them did not come from the lava itself but rather from air entangled in the lower lava flow or steam formed by the vaporizing of water contained in the moist surfaces over which the lava flowed. The various vesicles may be empty or filled with minerals. When the vesicles are filled with minerals, the basalt is said to be amygdaloidal.

The Watchung Basalt flows are famous for the beautiful zeolite minerals: prehnite, quartz, gypsum, thaumasite, datolite, apophyllite, pectolite, stevensite, and other associated minerals which have developed in the pillow basalts at the base of the upper flow unit of First Mountain in the vicinity of Paterson, in the area delineated as the former Lake Paterson. These deposits and their minerals have been described by many investigators but in particular, and more recently their mineralogy has been verified. This type of mineralization, but not in a pillow basalt environment has been developed on a much smaller scale elsewhere in First Mountain as at the Chimney Rock quarry near Bound Brook, N.J., and to a much lesser extent in the Second and Third Watchung Mountains.

Probable Character of the Pristine Gases

The composition of the pristine gases which were probably exsolved as bubbles into the liquid lava which later solidified as the Watchung Basalt may be approximated. The intrusive equivalents of the Watchung Basalt were examined by pumping the residual gases out of two Triassic diabases: one from New Market, Md., the other from Granton, Bergen Township, N.J. The unaltered diabase yielded a gaseous phase composed of 90 percent water vapor, slightly less than 5 percent H_2, 2 percent of S_2, 2 percent of N_2, about 0.9 percent CO_2, about 0.3 percent F_2, about 0.2 percent Cl_2, about 0.02 percent CO, and a trace of argon. The total volatile content amounted to about 30 cc per gram which is the normal value found for plutonic rocks. This analysis of the gases is probably representative of the composition exsolved into the Watchung Basalt.

Freshly collected basalts from the Hawaiian Islands contained about 5 to 6 cc per gram, and this again is the normal value found for lavas. This value would be the probable gas content of the solidified Watchung Basalt. This suggests that basalts retain about one-fifth of the gaseous constituents retained by the contemporaneously intruded diabases. This decrease in gas content is chiefly accomplished by the degassing on extrusion of the basalts.

The presence of 2 percent of sulfur analysis may be somewhat surprising, but the recent studies have definitely confirmed this. The observation was proposed that the main vesicle-forming phases in deeper pillows are probably dominated by the compounds of sulfur and carbon.

History of Tectonic Subsidence and Volcanism in the Newark Basin

Structure of the Newark Basin at the close of the Triassic Period

The inherited geometry of the Triassic basins, in their several disconnected parts, is a significant structural problem that has been studied by various geologists. The geometric outline of the Triassic basin in Connecticut was thought on the basis of field studies, particularly of the Bernardston Formation of Silurian and Devonian (?) age, to have been determined in the Devonian Period.

The close geometric relation was recognized as existing between the Triassic basins, extending from Nova Scotia to South Carolina, and the earlier Paleozoic folds, but were concluded from further study of the geological and structural maps of the eastern part of North America that the line of Triassic troughs cuts across the main axes of the preexisting Paleozoic structure, and that these troughs represent a new start dynamically which disrupted the old structural lines.

Movement along the border fault has been estimated to be 18,000 feet (5.5 km) and as much as 30,000 feet (9.1 km). A detailed study of the Ramapo fault system in New York and New Jersey suggests that the fault has a complex tectonic ancestry perhaps dating from Late Precambrian time and the relationship of the large intrusive Palisades phase to the east and the Watchung flows to the west of the basin suggests that the border fault at depth served as an avenue of magma ascent, although magma apparently never reached the surface along the fault. Contrarily, it is considered unlikely that the Triassic sediments extended much further northeast than the present exposure of the Triassic unconformity.

For the Triassic basin of New Jersey, the inherited geometry has been interpreted as occurring before deformation took place. The gross system in the graben consisted essentially of a stratiform mass made up of sediments, the three Watchung Basalt flow sequences, and included near its base, the discordant Palisade intrusive.

The layers of the stratiform mass were essentially flat lying. This mass was contained between the walls of the horsts. The dimensions of the graben based on estimates of about 30,000 feet (9.1 km) deep (approx. 6 miles) and 50 to 30 miles (80.5-48 km) wide. Geometrically the mass was bordered by marginal faults which dipped under the graben at 70°-75° and thus formed a trapezoidal block which narrowed with depth.

The geometry of the outcrop areas in Connecticut recognized the importance of the preservation of the Triassic rocks in the Pomperaug valley, an isolated outcrop area surrounded by granitic rocks and approximately 15 miles (24 km) west of the principal outcrop area of the Triassic of Connecticut. The general tilt in the Triassic belt appears to be the result of broad regional uplift in post-Triassic time, with maximum elevation along a north-south axis in western Connecticut. In northern New Jersey and southeastern New York the Triassic strata dip to the west, on the opposite flank of the regional arch. The Pomperaug area of western Connecticut, preserved from erosion by down faulting, indicates that typical Triassic sediments, with included basalt flows, existed on the crest of this arch, although they may not have been continuous between Connecticut and New Jersey. The presence of the Triassic rocks of the Pomperaug valley near the crest of the postulated arch is a key point in the argument for the former extent of the Triassic sedimentation and thus of the geometry of the Newark basin.

On the basis of the presently existing geometrical relations of the Triassic outcrops of New Jersey, Connecticut, and Massachusetts, it has been generally assumed that a suitable structural process which could bring about the present configuration in the Newark basin from Connecticut to Pennsylvania would require:

1. during the close of the Triassic Period or during the Jurassic Period, the Newark basin was subjected to a large deformation.
2. that the period of deformation was of long duration-perhaps slow movement over a long period of geologic time.
3. that the obvious longitudinal symmetry of the Newark basin in Connecticut and the Newark basin in New Jersey with strata and lava flow sequence dipping symmetrically in opposite directions indicated uplift over a longitudinal arch.
4. that the "border fault" accompanying the deformation is a normal fault and is indicative of tension in the crust.
5. that compressional forces were ultimately involved in forming the asymmetric syncline abutting against the fault scarp of the Precambrian granitic rocks.

The formation of a wide mountain arch under the Triassic graben was further extended. The Triassic block faulting, over the arch, illustrated the association of normal faulting with an extended tectonic area. The production of the arch brought about fracturing and settling along the walls and that these processes took place either during the arching or shortly thereafter. The later point is of considerable importance to the study of jointing in the Newark basin.

Structure of the Newark Basin After the Post Triassic Deformation

The geometry of the deformed graben in New Jersey cannot be described in detail for much of its contents have been removed. The angular relationships of the structure are the best preserved. The eroded remnant of this Triassic stratiform mass as observed today in New Jersey is tilted about 15°NW and strikes northeast. The marginal fault surfaces which originally dipped at 70°-75° under the graben were rotated 15°-20° and now dip 55°-60° in many localities. The northwest horst block of Precambrian granitic rocks, the Reading prong is topographically higher than the Triassic rocks, and the boundary between them is a large normal fault. The Triassic rocks form the downthrown block. The dip of the fault is to the southeast (**Figure 8**).

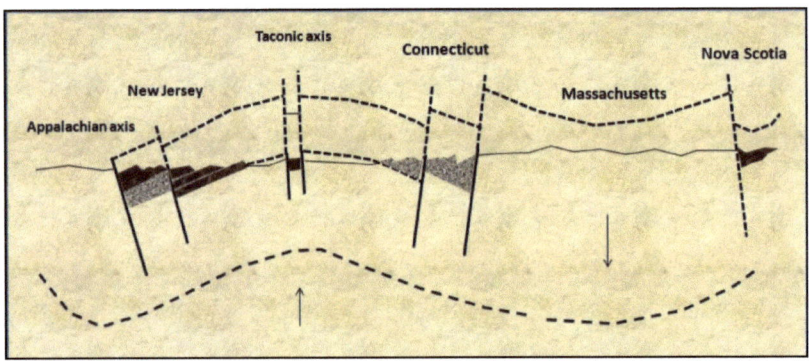

Figure 8. Generalized cross section of the Triassic of northeastern North America showing the postulated arching of the crust.

The linear dimensions and the volume of the structure, in contrast, are difficult to estimate. In the post-Newark faulting the cumulative movement along faults amounted to many thousands of feet. In the Delaware River Valley, the measured section of Triassic rocks from the base of the Stockton Formation to the great diabase sill at Haycock Mountain was found to be about 12,000 feet (3,658 m) thick. All of this section is below the Watchung lava flows, and thus it must represent the lower part of the Upper Triassic.

The border fault from Boonton to Stony Point truncates more than 10,000 feet (3,048 m) of Triassic rocks. A thickness of 30,000 feet (9,144 m) was given for the section in New York and New Jersey.

The suggested shape of the former structure at the close of the period of deformation is that of an arch. Approximately 22 miles (35.4 km) east of the west wall of the graben, on the east side of Trenton, N.J., the Triassic sediments have been exposed lying on the granitic rocks, there the Baltimore Gneiss. The contact dips 5°NW, parallel to the dip of the overlying Stockton sedimentary rocks. As the section is traversed eastward in wells, the granitic rocks are continuously found to lie under the Coastal Plain sediments and their appearance in the well logs deepens to the east, but little is known of the contact of the Newark rocks and the granite. The southeastern horst block of the graben, the Trenton prong, has been largely eroded away. It is not exposed in New Jersey, and its remnant, as revealed in well drilling, lies under the Coastal Plain deposits of Cretaceous age.

The geometry of the remnants of the arch in New Jersey, Connecticut, and Massachusetts suggest that its former geometric shape approximated that of a section of a right cylinder. This broad semi-cylindrical fold was bounded on both sides by the walls of the horsts, and the hinge line of the fold was parallel to the walls. The amount of stretching of the surface layers of such a buckled structure was estimated by calculating the length of an equivalent arc. The neutral surface was halfway between the bottom and top of the stratiform mass, that this section was unaltered in length, and that the angle of dip was the same on both sides.

Admittedly, this calculated increase in length due to buckling is a simplified treatment, but it gives us an approximate value for the amount of stretching of the rocks. This extension may be compared with that permitted by the tensile strength of the rocks.

From Hooke's law we have the relation that the strain produced in a body is proportional to the stress applied to it and that it is represented by the equation $E =$ stress/strain where E is Young's modulus. In a tensile strength test on a brittle substance the material is subjected to increasingly higher stresses producing increasingly greater strains until eventually the material fractures. Thus the stress required to rupture the material, the tensile stress, gives us a measure of the strain and therefore of the elongation at fracture.

The tensile strength of the stratiform mass in the graben may be estimated In the following manner. It will be assumed (1) that the composite stratiform mass behaves as a unit, and (2) that for the sedimentary rocks, the proportion of 43 percent sandstone and 57 percent shale is a reasonable distribution. Using the elastic and strength data, other elastic data, and the thickness data for the sedimentary rocks, lava, and diabase in the stratigraphic columns, the average tensile strength is estimated to be 340 lb/in2 (24 kg/cm2) and the average value of Young's modulus to be 2.7 X 10° lb/in2 (1.89 X 10^5 kg/cm2). This amounts to an extension of 243 inches (21 ft) or 21 feet for a 30 mile width.

A comparison of the estimated increase in length, by stretching, of the Newark rocks in the formation of the arch of 0.35 mile for a graben thickness of 10,560 feet with the estimated elastic extension of 21 feet in 30 miles, for a flat-lying stratiform block, shows clearly that, if the composite assemblage behaved as a single unit, rupture would take place with the formation of tensional cracks. Tensional joints would be developed. Actually, in arching, the graben assemblage did not behave completely as a unit, but slipping along the bedding planes must have occurred and tensional cracks and joints must have been formed.

As the upper surface of the arch developed tensional cracks, the neutral stress surface would shift downwards toward the compression side. Because the flank of the central anticline is no longer available for observation in New Jersey, this mechanical behavior cannot be proven. For the complementary case of the border syncline, where folding was produced by compression, field observation close to the fault wall has revealed small drag folds in the shale member of the Brunswick Formation. These appear in the shale beds that are confined by the more competent sandstone layers. These sandstone layers have developed a type of jointing similar to the ball and socket jointing that is sometimes observed in columnar basalt.

Regional Arch Formation

The formation of the regional arch by uplift from below may be treated as a problem in the mechanics of buckling. When a system is buckled, it changes its geometry; for masses involving large geologic structures, this change may be very significant. The geometry of the change which can take place, a slip surface must develop and that most commonly it is the bedding plane. The type of space problems that result are ideally developed at the ends of such a fold.

The geometric change in a folded slab will depend on the ratio of the length to the thickness of the slab and the curvature of the arch.

The estimated geometric change in the Newark graben will depend on the thickness assumed for the Triassic section, either 10,000 or 25,000 feet (3,048 or 7,620 m) and on the width of the basin. The estimated width of the basin in New Jersey is 50 to 70 miles (80.5-113 km) across. Field observations show that the preexisting fault .plane dipped 70° to 75° under the graben and thus gave rise to a trapezoidal volume (a truncated pyramid with its base at the Earth's surface). Field observations further show that during the arching of the basin, a major tension fault developed between the horst and the basin and that the displacement of these elements was of a considerable order of magnitude.

An arch may be produced in this manner by a number of mechanical systems. One explanation assumes that compressive forces were exerted by the granitic walls of the horst upon the layered sequences of the graben in the manner of a vise with slipping jaws. Another explanation assumes a large heat source to underlie the structure. The rocks above the heat source expand and form an arch. Another mechanism by which arching may take place is if a region is under tension the weakest structural element may be subject to dislocation. If this element is dropped into a zone where the element is forced to adjust to a new geometry, arching may occur.

If that part of the Earth's crust containing the Newark graben and its structural host, the granitic horsts, were simultaneously subjected to a strong tensional force acting normal to the long axis of the graben, then major rupture with dislocation would occur in the weakest part or parts. If such rupture did take place, then the field evidence should show where it occurred. The graben was not serially shear faulted. The only evidence of a major faulting is that along the graben-horst wall. This observation suggests that the weakest zones in the system were the preexisting fault planes between the graben and the horst walls, and rupture and dislocation took place in these zones with the contents of the graben dropping deeper into the opening created by the pulling apart of the walls of the horsts in response to the tensional force.

This dislocation of the graben probably took place over some period of time, and it was a discontinuous process. As a consequence of the adjustment of the down faulted block to the limitations of space in its new position, it folded into a central major anticline and two border minor synclines.

If, after the forming of this complex structure, erosion takes place to remove much of the central anticline, the border synclines will become topographically prominent features, especially if there are resistant layers such as basalt in the layered sequence which compose them.

The present topographic expression of the basalt flows of Third Mountain in New Jersey and its counterparts in Connecticut and Massachusetts fit this model. Long Hill, Riker's Hill, Hook Mountain, and Packanack Mountain are continuous parts of the southeast wall of the syncline. The New Vernon trap sheet and two small outcrops against the granitic wall in the vicinity of Towaco on Hook Mountain are parts of the upturned edge of the northwest wall of the syncline which are pressed against the granitic wall of the horst. The structure of the three Watchung Mountains conforms to the structural model proposed in **Figure 9**.

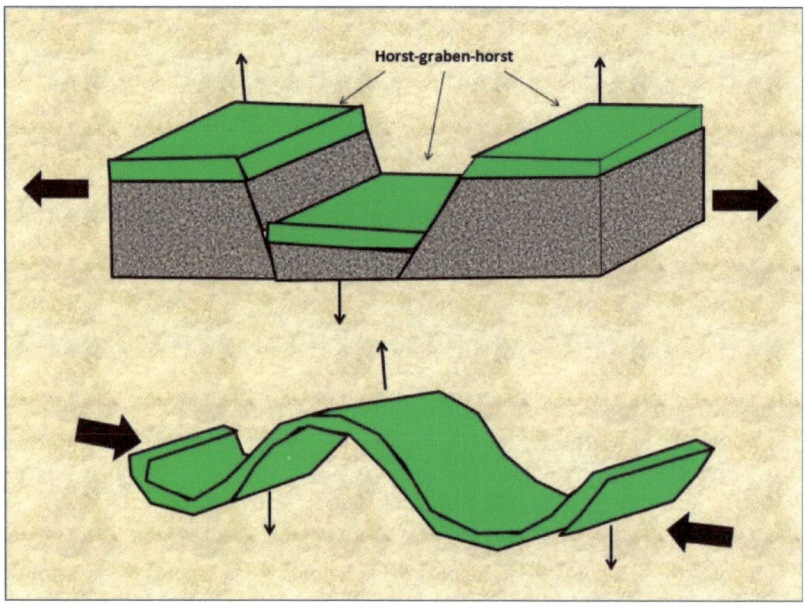

Figure 9. Schematic representation of a tectonic graben (top) and folds (bottom). Wide black arrows = horizontal tension (stretching); thin solid arrows = vertical relative movement; angled arrows = relative movement on the graben fault border. Bottom- Two synclines and one anticline; side black arrows = horizontal compression; narrow solid arrows vertical relative movement.

Triassic Tensional History

Development Effects of the Newark Basin and its Global Cause

The history of the Newark basin strongly manifests a state of tension in the earth's crust in the vicinity of eastern North America during Late Triassic and probably extending into Early Jurassic time. Three events in particular signalize tension as the major factor in the geologic forces which prevailed in determining Triassic events. These were rifting with the development of the rift zones and the formation of the graben, deep rifting in the graben which permitted volcanism with the outpourings of the flows and the intrusion of the diabase sills, and finally down faulting of the contents of the graben with the deformation of the block into an arch with border synclines.

The oft-repeated cycle of down faulting followed by basin filling suggests a region in a state of isostatic imbalance constantly striving to achieve balance. As a result of the gravitative adjustments in the graben, the sialic roots of the crust in this region were being forced deeper and deeper into the underlying sima, and it in turn was being laterally displaced. Near the end of Triassic time major deep rifting took place throughout the Newark basin. These rifts were multiple and long and the rift zones extended from Nova Scotia to at least as far south as North Carolina. They were also deep-seated rifts, for they tapped locally molten basaltic layers that were probably located in the vicinity of the Mohorovicic discontinuity at a depth of about 20.5 miles (33 km). The multiple character of the volcanicity proves that the rifting was not a single event. It probably consisted of three major periods of rifting, each period in itself being a multiple event. The time relations for these events in the Watchung have already been given.

In seeking a cause for the tensional processes operative in the Newark basin during the Triassic and probably into the Jurassic, a parallelism between the results of these processes and the major disruptive events involving the breakup of the continent Pangaea is clearly apparent. It is suggested here (1) that the period of graben formation in the Newark basin is a reflection of the state of tensional stress in the continent Pangaea at that time, and (2) that the development of the rift zones which permitted the outpouring of basaltic magma and the major event to affect the Newark basin (down faulting) occurred during the events leading up to and including the breaking apart of the landmass formerly composed of the then new continents of North America, Europe, and Africa.

It should be noted the connection between the voluminous outpourings of basalt in many regions of the Earth and the timing of continental drift. The view on this subject of volcanism is now generally accepted.

The suggestions proposed in this section concerning the events in the Triassic of eastern North America agree with the interpretations of geophysicists on the subject of the breakup of Pangaea and the formation of the North Atlantic. On the basis of paleo-latitude studies, conclusions were drawn that the rupture of the block and the opening of the North Atlantic did not follow each other closely in time. In fact, the block could have ruptured without creating an appreciable longitudinal opening, as the paleo-magnetic data presented here seem to indicate, and the best estimate is that the time of rupture is pre-Upper Triassic and that the time of the opening of the North Atlantic was probably Jurassic.

The rift that formed the Atlantic developed before the start of the Cretaceous about 120 million years ago. These data from paleo-latitude studies-rupture, and some separation-would correspond to the periods of strong tensional stress in the history of the Newark basin.

Supplementary Topics

Fissure Eruption Characteristics

Fissure eruptions are one of the important types of volcanic phenomena and are so-named because they originate from fissures-long cracks rending the surface of the earth. The lava ascends to the surface along the crack or system of innumerable cracks and wells forth for the most part as a quiet steady stream or sheet. Still more tremendous than fissure eruptions on steep slopes are such eruptions in the open country where the whole fissure becomes a volcano. The term "fissure eruption" as used in this case thus refers to those eruptions which take place from fissures of great length, commonly many miles long, and is not concerned with flank eruptions, or fissure eruptions on shield volcanoes. The considerable fluidity of the lava that issues from fissure eruptions has attracted much notice and has been attributed to their chemistry and volatile content. The observation that these lavas are exceptionally fluid at the time of extrusion, are attributed to the low viscosity to their chemical composition-they contain a notably high percentage of iron oxides, especially ferrous oxide. This opinion is based on the well-known fluxing action of ferrous oxide and manganous oxide in metallurgical slags.

The field observation that fissure eruptions, in general, show slight explosive activity can be concluded that the magmas were not rich in a gaseous phase. It has also been suggested that the fluidity of the magmas giving rise to fissure eruptions has been maintained because of the rapid rate at which the magma moves from its chamber to the Earth's surface. If this is true, then the ascent of lava on a system of innumerable fissures would greatly influence the maintenance of the fluidity and by providing many paths of travel to the surface would allow a large volume of molten rock to be moved quickly.

The eruption and flow rate of basalt from fissure eruptions were studied with particular reference to the basalts of the Columbia River Group. The probable total time of emplacement of such flows was calculated and determined that, for example, the 3 m conduit with 3 km strike length gave a total flow rate of about 0.6 km3/hour for a hydraulic slope of 0.1. A reference volume of 100 km3 for a 'typical' Yakima flow would then imply a total emplacement time of about a week (2 days for the 5 m conduit).

The possibility of even slight explosive activity initiating a fissure eruption has been questioned on the basis of recent field studies. The tephra produced at the Eldj a fissure eruption in Iceland resulted mainly from a very vigorous lava fountain activity and not from explosive activity. The geometry of the surface formed by the intersection of a fissure and the surface of the Earth would be similar to that of a long ribbon, and this surface would have a tremendous area. Lava welling forth from this vent and presented, at every instant, with such a large surface would be capable of discharging any significant excess of gaseous phase, and thus there would be no tendency to buildup explosive energy.

Geological descriptions of the fissure eruptions in the Columbia River plateau, in Britain, the Deccan of India, in Iceland, and elsewhere indicate that a fissure eruption proceeds as follows: The fissure or fissures form, followed later by very minor explosive activity, and then the molten lava pours out of the fissure system as a great flood. The lava fills all the depressions in the land surface and levels the topography which it covers in a manner analogous to water flooding. It is possible in examining a lava field of some recent fissure eruptions to actually map the flow lines on the surface of the lava sheet as was done at Mount Eccles. If the eruption takes place over a flat land-surface, the flood pours out over vast areas, and the flows may attain great thicknesses.

There may be a succession of flows so that one sheet lies directly above another, or the flows may alternate with sediments deposited during an interruption in the volcanism. During an eruption, the fissures may lengthen. If a fissure is closed with solidified lava, it may reopen to permit the flow of lava again. This passage of several flows of lava through the same fissure gives rise to multiple dikes. Old fissures may be paralleled by new fissures from which the lava flows. Some lava-filled fissures never opened at the surface. Regarding the fissures observed, these fissures whether due to sudden shocks or slow disruption, were produced with such irresistible force as to preserve their linear character and parallelism through rocks of the most diverse nature, and even across old dislocations having a throw of many thousand feet. Yet so steadily and equably did the fissuring proceed over this enormous area that comparatively seldom was there any vertical displacement to the sides. We rarely met with a fissure which had been made a true fault with an up thrown and down thrown side.

The closing stage of a fissure eruption was interpreted to be indicated by the presence of cones along the fissures, which merely mark the sites of the feeble final spasms of activity. It is obvious that the preservation of these fragile cones, almost intact after tremendous volumes of lava have poured out, 3 cubic miles (12.5 km3) at Laki in 1783, is testimony to their very late development in the dying stages of the eruption. The closing phase of the fissure eruption at Mount Eccles in Victoria, Australia, is illustrated where a line of small cones traces the position of the fissure from which welled the large volume of lava. It is also apparent from these illustrations that, owing to the vast areal extent of fissure eruptions, the detection of the fissures may be very difficult. This is particularly true in an area whose topography was relatively flat previous to volcanism and where erosion has removed the surficial evidence of the fissure. The field appearance of a filled fissure now exposed as a dike in a tributary to the Salmon River in western Idaho, that the formation of this dike marked the closing phase of a fissure eruption.

The shape of a fissure eruption in open country, where it is unlimited by topographic barriers, tends to be pseudo-rectangular with its longest dimension parallel to the fissure system. This is illustrated in the schematic diagram given in **Figure 10**. It is for this geometric reason that fissure eruptions are called linear eruptions by some petrologists and volcanologists. The concept of lava flowing out in the manner of a flood had been envisioned by a number of geologists.

Figure 10. Schematic representation of a fissure eruption showing the linear formation of the late stage vents in the closing phase (top and middle) and after the eruption has ceased (bottom).

In explorations of the Colorado River, an observation was that the fissure of this fault had been the channel through which floods of lava have been forced from depths below into the upper world. The history of the Tushar explained while the grander floods poured out over everything and spread out over great expanses of the mountainside. "Flood basalt" was later used to designate the rocks formed by a fissure eruption in allusion to the manner of emplacement of the lavas. Although such rocks have also been called plateau basalts, because some plateaus are formed of these rocks, the occurrence of these lavas as plateaus is due solely to geologic changes unrelated to volcanism.

Fissure eruptions were also called "pedionites", based on the Greek word pedion meaning flat. Such eruptions were regarded as producing the least limited type of volcano. The actual time span of fissure eruptions and their geometric extent are known for only a few of these events. Fissure eruptions were first called massive eruptions, and it is of interest that the first recognition of the importance of fissure eruptions was made on the Pacific coast in 1868.

This work was ridiculed in England and did not find favor in Europe until 1881-82 when after return from field study in the western United States affirmed the field observations on massive eruptions exposed the error. Extensive geomorphological studies of the volcanoes in the Auvergne region of France observed the results of the erosion of the cones and craters of central type volcanoes; on this basis a conclusion was reached through examination of the erosional remnants of the former loci of volcanism.

Lava flows in the British Isles, specifically in the basaltic plateaus of Ireland and Scotland, where the abundant related dike swarms had no connection whatsoever with any central type vents. In the Columbia River plateau, extensive studies in the area were described in conclusion as follows: A recent journey in Western America has at last lifted the mist from geological vision. Some of the lava-fields of the Pacific slope were examined, and the realization that the conditions of volcanism, the reality of the distinction between 'massive eruptions' and ordinary volcanoes with cones and craters was made.

Bubble Formation Processes and Movement

Vesiculation is the process by which a very viscous liquid such as a basaltic magma exsolves a gaseous phase to form bubbles. The interfaces of these bubbles with the liquid are commonly spheroidal. Their former presence in the magma is attested to by the presence of cavities in the basalts called vesicles. The formation of bubbles in silicate melts has been studied extensively by glass technologists. To a lesser extent, it has been investigated by silicate chemists concerned with metallurgical slags. Their experimental studies and particularly their large-scale industrial plant practices show that: 1. if a molten silicate melt flows over a rough surface bubbles will form. When the melt wets the surface, bubbles are nucleated by the air entrapped between the melt and the rough surfaces. 2. Large bubbles will move more rapidly than small bubbles-this follows from a consideration of Stoke's law. 3. A decrease in viscosity of the melt will increase the rising velocity. The rising velocity gets smaller and the velocity increases. 4. As a bubble rises through zones of super-saturation, it grows by diffusion of gases through the melt, and, as its radius increases, its rising velocity increases. The converse situation applies if the bubble enters zones of under-saturation. 5. Bubbles that reach the surface of the melt may burst, this will depend on the thickness of the film of glass containing the bubble, the internal pressure of the bubble, and the surface tension of the melt.

It had formerly been assumed that the large bubbles grew chiefly through the coalescence of small bubbles as they collided on their ascent in the melt and that the larger bubbles served as collectors of small bubbles which they then dragged along with them. Modern studies show that this mechanism plays only a minor part in the growth of bubbles. Bubbles grow by diffusion of gases from the melt at the level where the bubble temporarily rests. As the bubble moves to higher levels in the melt, more gases diffuse to it and its size increases; this growth goes on until the bubble reaches the surface of the melt or another interface.

Experimental studies on glasses indicate that thermal shock may aid in the nucleation of bubbles. A system is said to be in thermal shock when it, or any part of it, is suddenly subjected to a drastic change in temperature.

Natural Lava Flow Bubble Formation

The basalt magma, when it is in the earth before extrusion will have its internal pressure (the vapor pressure) at a much lower value than that of the external pressure. As the magma ascends to the surface, its vapor pressure exceeds the external pressure upon it (sum of hydraulic pressure of overlying lava and the atmosphere), gases will be exsolved to form bubbles. As the lava continues to flow over the rough and porous sandy floors, this movement could supply considerable trapped air and steam which could be incorporated as bubbles in the flow.

The process of bubble formation and movement is thus regulated by two phenomena-the equilibrium process and the rate processes. The equilibrium process is controlled by the vapor pressure of the melt. If, as during a volcanic eruption, the external pressure is reduced then vesiculation of the lava can take place. The vapor pressure of a melt is generally not constant, and it can be increased by crystallization of volatile free silicates. The rate processes operating in a melt are nucleation of bubbles, gas diffusion to the bubbles, and the rise of the bubbles in the melt.

The addition of bubbles at the base of the flow when it pours out over rough surfaces is an aid to the rate processes in that it probably provides nuclei. As the bubbles ascend in the liquid part of the flow their shape may vary, and this characteristic will be dependent upon a number of factors but the most important will be the velocity with which they ascend.

After the flow has come to rest, it continues to evolve bubbles and begins to cool. The natural system is complicated by the fact that the system is losing a gaseous phase to the atmosphere in significant amounts only so long as the surface is still liquid. After the surface of the flow hardens, the loss by escape into the atmosphere is greatly decreased and is probably chiefly limited to escape along the joint systems in the vesicular tops of the flow. After the vesicular top has formed to a depth of a few feet, the bubbles that rise to the liquid-solid interface at the base of the hardened top will accumulate there.

The evolution of bubbles in the lava flow as a result of the exsolution, after it has been emplaced, is a continuous process, but the rate of bubble formation may be changed significantly by a number of other physical factors. These factors may operate so as to produce unusual concentrations of bubbles in the later stages of solidification of the lava flow. As the lava flow continues to solidify the residual liquor will be richer in volatiles, and these will decrease the viscosity thereby aiding the formation and growth of bubbles. Concomitant with these conditions there will be developed a greater tendency for the system to form larger crystaJs.

Some of the bubbles which migrate to the upper surface of the liquid flow will burst and release the gaseous phase to the atmosphere. The pressure in a bubble is the sum of three pressures, the atmospheric pressure, hydrostatic pressure, and the pressure caused by surface tension. The internal pressure within the bubbles of various sizes caused by the surface tension can be calculated. The surface tension of the Watchung Basalt has been calculated using the average composition of 22 specimens, the surface tension factors for the oxides, and taking the factor for FeO as the same as MgO. This gives a value of 405 dyn/cm for the surface tension at 1,200°C. Using the formula for the internal pressure at 1,200°C for bubbles of 1/1,000-mm, 1-mm, and 6-mm radius, the respective pressures are 8.1, 0.0081 and 0.00135 atmospheres. These values show that the pressure caused by surface tension in the liquid lava of the Watchung flows becomes negligible for bubbles 1 mm in radius and larger compared to the atmospheric pressure and the hydrostatic pressure.

It is apparent that the pressure to form a new bubble is tremendously large, one can therefore assume that one is dealing here primarily with a spontaneous growth of seed nuclei due to 'thermal fluctuations' and that these nuclei can continue to grow later on.

Those bubbles that do not burst will retain their gas, and this gaseous phase will probably react with the walls of the vesicles as the temperature drops in the system. Some of the gas may re-dissolve in the melt. Any gas not used up in such reactions will later escape in the process of weathering.

References

Faust, G.T., 1975. Review and Interpretation of the Geologic Setting of the Watchung Basalt Flows, New Jersey. Studies on the Watchung Basalt Flows of New Jersey. Geological Survey Professional Paper 864-A. United State Geological Survey, US Department of the Interior and references contained within.

Wikipedia, 2019. The Palisades Sill. Posted on the internet.

Wikipedia, 2019. Watchung Mountains. Posted on the internet.

____. Central Atlantic Magmatic Province (CAMP): The Palisade Connection. Chapter 5. Undocumented publication posted on the internet including references within.

www.ingramcontent.com/pod-product-compliance
Lightning Source LLC
Chambersburg PA
CBHW040233220526
45473CB00001B/219